数字化
应急医院设计及
建造技术

Design and Construction Technology of Digital Emergency Hospital

中南建筑设计院股份有限公司　编著

中国建筑工业出版社

编委会

序

雷神山医院、方舱医院的建设生动体现了中国力量、中国精神，集中诠释了中国效率、中国技术，高度彰显了中南院人同舟共济、守望相助的家国情怀。

一、非常之时，非常之事

2020年农历新年，为应对蔓延的新冠肺炎疫情，武汉市决定迅速建成火神山医院、雷神山医院。除夕之夜，中南建筑设计院股份有限公司（以下简称"中南院"）承接了雷神山医院的设计任务，董事长李霆挂帅，总经理杨剑华一线指挥，当晚组建设计团队，赶赴现场踏勘，连夜开始各专业设计。当时是病毒传播高峰期，中南院人逆势而行，迎难而上。之后，中南院又承担了武昌方舱医院、江夏方舱医院、江汉方舱医院等几十家湖北省及外省的方舱医院和应急医院的改建任务。据统计，在武汉市建成并投入使用的方舱医院中，由中南院承担设计任务的达40%。中南院人在非常之时，挺身而出，共克时艰，行非常之事，做非常之人。

雷神山医院的设计是在特殊时期、特殊背景下展开的，时间紧、疫情重、压力大、要求高，全民期待、举世瞩目。应急医院是救命工程，与时间赛跑，与"死神"较量，每一个细节都必须精益求精，每一张图纸都必须精确无误。设计师们没有辜负国人的期待，没有辜负市民的信任。他们夜以继日，以严谨的作风、顽强的意志、忘我的精神、无私的奉献，在短短几天时间内完成了通常需要半年时间才能完成的工作，如期拿出了施工图，主动配合现场施工，确保了工程顺利完工。

二、非常之技，非常之能

建非常之功，必须拥有高超的技术水平和解决复杂问题的应变能力，平静经历风雨，从容应对挑战。尽管是设计临时应急医院，但是设计师们以极其严格的标准和高度负责的态度进行设计，是非常之作。可以说，技术水平、奉献精神、创新能力是制胜的法宝。

雷神山医院是数字化设计的产物，是技术创新的范例。在几天时间内设计一座建筑面积近8万平方米的大规模集装箱医院，没有先例可以借鉴，需要全新探索。设计方案必须解决三大难题：快速建造并立即启用，防止对环境造成污染，避免医护人员感染。为此，设计遵循标准化、模块化、装配式原则，满足极其紧张的工期要求，实现安全感染控制和快速建造。设计师采用BIM（Building Information Modeling，建筑信息模型）技术建立数字模型，考虑装配式建筑特点，为施工安装快速完成提供便利。此外，远程办公、网上协同设计、数字化图纸交换等信息化技术得到全面应用，为建筑设计行业管理转型升级积累了经验，提供了实战案例。

雷神山医院是多专业设计创新的具体演示和智能化集成。建筑、结构、机电、暖通、给水排水设计都需要先进的设计理念和深厚的技术实力作支撑，各专业设计师巧妙地解决了一系列应急医院设计建造面临的技术问题，如"三区两通道"布局优化、基础和结构快速施工、机电设备智能化、通风排风系统数字仿真、排水安全处理、应急工程投资管理等。设计全过程充分应用了信息化手段，达到了智能化效果。

三、非常之心　非常之情

勠力同心，众志成城。在全民抗疫之际，建筑界同行和协作单位包括外国友人主动与中南院联系，提供技术支持。构建 BIM 平台、开展 CFD（Computational Fluid Dynamics，计算流体力学）仿真、启动实地监测、讨论成果总结、进行学术交流，这一切都体现了国内外同行的爱心和善举。因为特殊的项目结缘，因为技术的协作相遇，这是大爱无疆的深刻记忆。

中南院历来重视与国内外同行合作设计和研发，这次也不例外。清华大学的团队主动参与，无私协助，提出临时医院排风环境影响的快速模拟方法，以开源 CFD 软件为基础，为快速分析提供了专门工具。法国达索系统公司协助建立 BIM 模型，运用 XFlow 软件对气体组织和污染物扩散进行模拟分析，提供了一套可靠的 CFD 仿真验证及优化设计方案，有效避免院内交叉感染及医院对周边环境的污染。以设计成果为基础，设计团队一起研讨新技术新应用，在权威工程力学杂志上发表论文，为提升医疗建筑数字化设计水平提供了全新参考。

一批年轻设计师在袁培煌、李霆、桂学文等大师和众多技术专家的精心指导下，在严苛的环境中得到锻炼，脱颖而出，成为技术骨干。中南院多年来一直注重前沿新技术研发和年轻技术人才培养，特别重视数字化和智能化技术应用，成立专门的研究机构和设计中心，率领年轻一代不断探索、不断创新，为中南院技术进步培养人才、

储备能量。前辈们言传身教，以自身的修为引导、激励年轻人的成长。青年才俊在危难时刻勇挑重担，敢于创新，不负重托，不辱使命。

四、非常之感，非常之言

山河无恙，人间皆安。经过国家和各界人士三个多月的鏖战，疫情得到基本控制。后疫情时代留给我们无限思考。经历了争分夺秒与病魔抗争的非常之时，抒发居安思危的非常之言。

2020 年 4 月 5 日，方舱医院这一中国创举的主要倡导者王辰院士在中南院发起了方舱医院建造、管理与发展专题研讨会，共同复盘生命方舟如何不负众望。研讨会从多角度分析了方舱医院在疫情期间的作用、特点、建设的重点、难点，对进一步提升方舱医院的建设和管理水平意义深远，为未来在"平疫结合"思路下进行应急医院的改造、大型场馆的设计提供了极其宝贵的经验。研讨会上，王辰院士指出这种成果的背后是政府的高效统筹，是企业家的倾囊相助，是建设者的努力付出，是老百姓的积极配合，是患者的绝对信任，是多方付出得来的。

中南院人在国家需要的非常之时，行非常之事；应抗疫需要的非常之技，展非常之能；怀人民需要的非常之心，呈非常之情；以生死较量的非常之感，发未雨绸缪的非常之言。中南院人就是这般英勇无畏、前赴后继、义无反顾地投入到这场无声的战"疫"中。可敬可爱的中南院人身上，彰显了独特的基因。

一是责任基因。一声令下，设计师们主动请缨，这种忘我的牺牲精神，是源于深入血液、刻

入骨髓的责任感，源于对国家和人民的责任感，源于强烈的人类命运共同体意识，这驱使他们在危难面前毫不退缩、迎难而上。

二是创新基因。中南院素来秉承"创新创意，至诚至精"的企业理念，技术创新是企业的生命。临危不乱、自信应对，源于扎实的技术积淀。技术水平、创新能力是制胜的法宝，代代传承的创新基因，让中南院人在非常时刻迸发出强大的创造力。

三是开放基因。多年以来加强国际合作、深耕国际业务，铸就中南院多元包容的开放基因，兼收并蓄、博采众长，拓展国际视野，提升设计品质。

雷神山医院和方舱医院寄托着降伏病魔的美好愿望。随着武汉市疫情防控的成功，雷神山医院、方舱医院完成其特殊使命，尘封于历史。中南院为养育自己的这座城市度过劫难、实现凤凰涅槃呕心沥血，以赤子之心报恩报国，感人至深。每一位参与项目设计的人都为有机会救人救难而自豪，感谢国家和人民的重托，雷神山医院和方舱医院是他们内心深处永恒的记忆，是他们精神上的期许之地、报恩之所、祈福之山。

雷神山医院和方舱医院是智慧的结晶，是心灵的呼唤，是奉献的升华。为铭记这段特殊而又难忘的历史，全面总结技术成果，及时分享抗疫工程设计经验，中南院决定编写本书。杨剑华总经理邀请我们为本书作序，我们欣然应诺，谨表对中南院抗疫逆行者的敬意。

王　辰

中国工程院副院长、院士

中国医学科学院北京协和医学院校长

张柏青

工学博士

湖北省政协副主席、中南设计集团董事长

FOREWORD

Leishenshan Hospital and other makeshift hospitals (Fangcang Shelter Hospitals) vividly demonstrate the power, spirit, efficiency and technologic progress of China. As the project designer, Central-South Architectural Design Institute Co., Ltd. (CSADI) participated in the projects with a sense of shared destiny and a sentiment toward family and country.

I. Prompt Actions in Critical Times

Around the Chinese New Year in 2020, Wuhan had to make a decision of building two emergency hospitals as quickly as possible, which were called Leishenshan Hospital and Huoshenshan Hospital, to receive more patients infected by COVID-19. On the Chinese New Year's Eve, CSADI set about designing Leishenshan Hospital. Li Ting, Chairman of the Board, led the task in person, and Yang Jianhua, General Manager, worked in front line as a team leader. Designers were immediately teamed up and went to site for survey and kicked off the design right away. It was the peak time of virus transmission then, so most people would try hard to keep away, but our design team faced it head-on. After designing Leishenshan Hospital, CSADI again engaged in designing dozens of makeshift hospitals and emergency hospitals in and out of Hubei Province, among which are Wuchang Makeshift Hospital, Jiangxia Makeshift

Hospital and Jianghan Makeshift Hospital. Statistics show that CSADI took up to 40% of makeshift hospital design in Wuhan. In such a critical period, CSADI staff made a bold decision and took unusual action to overcome hardship by working together.

As the design of Leishenshan Hospital was carried out in a special period, it faced so many challenges including tight schedules, severe pandemic situation, heavy pressure and high requirements. The project was catching attention from people at home and abroad. Emergency hospitals were born to save lives in high risk. Time must be saved and evil of death must be fought against. The design team must do their best in every detail and produce accurate drawings. Thankfully, the design team didn't let the Wuhan people down. With a strong will, they worked carefully and selflessly day and night and finally finished construction drawings. The whole process would usually take half a year. The designers actively worked in partnership with the construction contractor on site so that the project was able to be completed in time.

II. Combination of Unusual Technologies and Capabilities

An unusual project can not be completed without special technologies and extraordinary capabilities in dealing with complicated situation. Although the designers were dealing with emergency hospitals with temporary use only, they did it fantastically in strict compliance with the standard and with a high sense of responsibility. It can be concluded that they made it by combining technologies, devotion and innovation.

Leishenshan Hospital is a product of digitalized design and an example of applying innovative technologies.Designers had nothing to refer to in designing such a large scale container hospital of 80 000 m^2 within several days. They must remove three obstacles : rapid construction and handover, minimum environment pollution and virus transmission. Hence, the team decided to apply standardized, modularized and pre-fabricated structures to meet extremely tight schedule, safety control and rapid building. They applied building information modeling (BIM) technology to establish digital models and incorporate prefabrication to facilitate rapid construction. Besides, they gave full play to information technologies including remote office work, synchronized online design and digital drawing communication. These operations help build up case-based experience in transforming and upgrading architectural design and management.

Leishenshan Hospital is a demonstration of bringing multi-disciplined cooperation and is a result of intelligent integration. New ideas in design and technological capability are necessary in following disciplines including architecture,

structure, mechanics and electrics, heating and ventilation, as well as water supply and drainage. Designers in different disciplines have skillfully solved technological problems emerging from building emergency hospitals. These problems include but not limited to the layout optimization of three areas and dual passages, rapid construction of foundation and structures, intelligent buildings of electromechanical equipment, digital simulation of ventilation system, safe drainage treatment, and emergency project investment management. Information technologies were fully applied in the entire process of design to achieve intelligent building.

III. Special Thinking and Partnership

To fight a battle, we must put our heart and strength together. On the occasion of the nationwide anti-epidemic actions, CSADI received help and technological support from its counterparts in the construction industry and other partners from overseas. Love and kindness shown by the national and international fellows were seen in BIM platform conceiving, CFD simulation, on-site monitoring, summary of discussion results, and academic exchanges. They met and built a tie as a result of the special project and technological cooperation, which left them a deep memory of boundless love.

CSADI has always been attaching great importance to cooperation with domestic and foreign counterparts in design and R&D, and there is no exception this time. The team from Tsinghua University took the initiative to join them, and proposed a rapid simulation method for the impact of ventilation environment in the makeshift hospitals. This method, based on the open source CFD software, acted as a special tool for rapid analysis. The French Dassault Systèmes assisted them in the establishment of the EIM model, used XFlow software to simulate the gas organization and pollutant diffusion, and provided a set of reliable CFD simulation verification and optimization solutions to effectively prevent cross-infection within the hospital and from polluting the surrounding environment. Based on the design result, the design team discussed new technologies and applications together, and published papers in authoritative engineering mechanics magazines, providing a new reference for improving the digitalized design of medical facilities.

A group of young designers, under the careful guidance of Mr. Yuan Peihuang, Mr. Li Ting and Mr. Gui Xuewen, were trained in harsh environments. They have grown to be technological backbones. Mr. Li Ting has profound knowledge, broad vision and rich experience. Over the years, he has been focusing on the development of cutting-edge technologies and the training of young technical talents. He puts emphasis on the application of digital and intelligent technologies. For this purpose, special research institution and design center were established. He guides the younger generation to continue exploring and pursuing innovation, so as to train talents for CSADI to drive technological progress. He educates the young by his words and deeds and encourages them to research and develop. It was a good chance for young talents to take up their responsibilities, drive innovation and live up to people's expectations in such a hard moment.

IV. Special Feelings and Special Words

After more than three months of unremitting efforts of the state and people from all walks of life, the epidemic was basically controlled and life returned to normal. In April, 2020, Leishenshan Hospital and other makeshift hospitals were suspended. When we look back, special feelings well up. As one way to keep vigilant in the time of safety, we wrote this foreword. Wang Chen, Academician of Chinese Academy of Engineering, and main proposer of building makeshift hospitals, which was an inventive action in China, launched a workshop on construction and management of makeshift hospitals at CSADI on April 5, 2020. A topic question was raised in this workshop: how can makeshift hospitals live up to people's expectations? The role and characteristics of makeshift hospitals as well as the key points, and challenges of building them during the epidemic were analyzed from multiple aspects in the workshop. It has far-reaching significance for further improving the construction and management of makeshift hospitals. The workshop drew lessons and experience on designing and converting large-scale space into makeshift hospitals so that these hospitals serve their roles both during normal time and emergency period. At the workshop, Wang Chen pointed out that behind this victory was the efficient coordination of the government, the generous financial assistance from entrepreneurs, the constructors' great efforts, active cooperation of the people, and patients' absolute trust.

CSADI did display its extraordinary capabilities when it was needed in an unusual time and protected the city of Wuhan with an extraordinarily kind heart. The staff of CSADI cast themselves into this silent battle against the epidemic with bravery and fearlessness. The respectable and lovely CSADI staff showed their unique inner characteristics.

The first characteristic is sense of responsibility. Upon an order, designers stepped out and applied for participation. The spirit of sacrifice originated from the sense of responsibility that flows in their blood. It was a sense of responsibility to the country and the people, and a strong awareness of shared future, which drove them to face up the crisis without hesitation.

The second characteristic is innovation. CSADI has always been adhering to the cultural philosophy of "being innovative, creative, sincere and professional". It is acknowledged that the life of an enterprise rests on its technological innovation. It was the solid accumulation of professionalism and technologies that made CSADI keep calm and confident in dealing with the situation. High technological level and innovative ability were the magic weapon to the victory. Technology and innovation have been internalized in the spirit of CSADI staff of different generations, allowing them to burst out their powerful creativity at this critical moment.

The third characteristic is openness. Over the years, CSADI has been strengthening design cooperation with its international counterparts. This has been developed into an open attitude, enabling CSADI staff to learn from others, broaden their international horizon, and finally improve design quality.

Leishenshan Hospital and other makeshift hospitals in Wuhan carried people's hope for overcoming COVID-19. With the success of prevention and control of the epidemic in Wuhan, Leishenshan Hospital and other makeshift hospitals were closed and will be sealed into the history. The city of Wuhan nurtured CSADI. For return, CSADI has made painstaking efforts to help this city get through the hardship. It's a touching story. Everyone involved in the project design is proud of having the opportunity to save people's lives. CSADI is indebted to the country and the people. Leishenshan Hospital and other makeshift hospitals will be in CSADI's memory forever.

Leishenshan Hospital and other makeshift hospitals are the crystallization of wisdom. They were called out for people's need from the heart, and what they did is beyond their role as hospitals are owed to the dedication they witnessed.

To remember this special and unforgettable period, comprehensively summarize the technological achievements, and share the anti-epidemic experience in project design in a timely manner, CSADI, decided to have this book published. Yang Jianhua invited us to write a foreword for this book, and we were pleased to accept this invitation. For us, this is a way we can express our respects to the countermarching people of CSADI in the fight against the virus.

Wang Chen

Academician and Vice President of Chinese
Academy of Engineering,
and President of Chinese Academy of Medical
Sciences & Peking Union Medical College

Zhang Baiqing

PhD of Engineering, Vice Chairman of Hubei
Provincial Committee of the Chinese People's Political
Consultative Conference (CPPCC), and Chairman of
Central-Southern China Engineering Consulting and
Design Group Co., Ltd.

前　言

当前，有关新型冠状病毒的研究有了较大进展，但病毒何时能够被彻底消灭，对人类来说仍然是未知数。危机意味着危险，也昭示着解除危险的机遇和道路。有效应对这一场全球公共卫生危机，需要加强国际合作与交流，构建人类命运共同体。

2020年年初以来，中国举全国之力用三个多月的时间基本控制住国内疫情后，立即尽己所能向上百个国家和地区以及相关国际组织提供了紧急援助物资，并积极分享疫情防控经验，践行构建人类命运共同体的庄严承诺。在中国本土疫情防控中，雷神山医院、火神山医院两座临时应急医院及一批方舱医院的设计建设，对提升收治能力、快速控制疫情起到了非常重要的作用。

中南院建院于1952年，已有70年历史积淀，是中国知名建筑设计企业。中南院秉承"创新创意，至诚至精"的企业理念，70年来，始终坚持为社会提供优良的建筑设计作品和一流的工程技术服务，先后在国内29个省、自治区、直辖市及国外21个国家、地区完成了16000余项工程设计，其中有700余项工程获国家级、部级、省级优秀设计奖和科技进步奖。近年来，中南院大力推进转型升级，强化战略布局，不断推进专业化团队建设，先后成立了医疗健康事业部等多个产品事业部，在专业化领域不断做精做细。多年来，中南院在医疗建筑设计板块深耕细作，形成了完备的组织框架和技术支撑体系。截至目前，中南院已累计参与设计国内外200多个大型医院项目，包括国内顶级医院华中科技大学同济

医学院附属同济医院、华中科技大学同济医学院附属协和医院，还积极响应商务部号召，在非洲和拉美等地区援建了一批医院。医疗健康事业部在这次抗疫应急工程建设中展示出高超的专业技术水平和优良的统筹协调能力，从城市规划、建筑方案设计到全工种的施工图设计，都抢在了前面，为抗击疫情赢得了时间，为救助新型冠状病毒肺炎（简称"新冠肺炎"）患者提供了重要保障。在本次疫情防控中，中南院累计完成抗疫医院设计38个，其中雷神山医院等医院建造、改造设计14个，方舱医院改造设计21个，方舱医院EPC项目2个，指挥部改造设计1个，共计助力产生3万余床位。

雷神山医院是参照2003年抗击"非典"期间北京小汤山医院模式，集中收治新冠肺炎患者的医院。经过参建各方十余昼夜的全力奋战，2020年2月8日雷神山医院收治了首批患者，4月14日雷神山医院患者清零。雷神山医院采用模块化设计，主要包括隔离医疗区、医护生活区、保障功能区，建设用地面积约22万平方米，总建筑面积约8万平方米，整体按照相关标准设计。医院规划有医护生活区建筑约2.75万平方米、隔离医疗区及保障功能区建筑约5.25万平方米。隔离医疗区普通病房总床位数建设目标为1500张，重症加强护理病房（ICU）床位数为60张；医护生活区可容纳医护人员约2300人。中南院设计团队接到任务后12小时提交设计方案、3天完成施工图设计、10余天完成平常半年才能完成的设计工作，彰显了中国速度。项目因医院

净化系统级别在国内最高，得到国家卫生健康委专家的充分肯定，原北京小汤山医院院长、雷神山医院建设专家顾问张雁灵称赞"雷神山建设标准高于小汤山"。

方舱医院理念由中国工程院院士王辰提出。将一批体育馆、会展中心等改造为大规模集中救治轻症患者的临时医院，旨在让定点医院腾出床位收治重症患者，使得医疗资源得到有效配置。方舱医院理念提出后，迅速被政府采纳，要求第一批 3 座方舱医院 48 小时内改造完毕并收治患者。在武汉市建成并投入使用的 16 座方舱医院中，由中南院设计的达 6 座之多。据官方数据统计，从 2 月 5 日到 3 月 10 日，武汉市 16 座投入使用的方舱医院共收治 1.2 万余名患者。方舱医院是中国在抗击疫情关键时刻的关键之举和创举，短时间内实现了"应收尽收、应治尽治"，拧紧了疫情扩散的"水龙头"，对疫情遏制起到了至关重要的作用。

截至 2020 年 4 月底，武汉市疫情防控已经取得决定性成果，这些应急医院也已经完成它们的历史使命。中南院及时总结抗疫应急工程设计经验，编写本书，分享抗疫应急工程设计建设经验，以期促进国际合作，共同抗击疫情。

本书主要内容分为技术基础、技术创新两个部分，以雷神山医院和方舱医院为切入点，系统全面地总结数字化应急医院的建设经验。在这场战"疫"中，中南院团队完成的设计项目类型涵盖范围广，在很多方面都经受住了实践的检验，获得了珍贵的第一手资料。一方面，对于各类型的医疗建筑在总体规划和建筑设计方面有哪些问题需要厘清，在室内外环境控制方面有什么技术经验可以分享，在装配式防疫医院的设计与建造方面还有哪些创新手段，特别是如何利用 CFD 仿真模拟、BIM 三维建模等数字化技术最大限度地保障快速建造和减少感染，值得我们继续研究；另一方面，对于在紧急状况下临时改建方舱医院、利用特殊场所临时改造隔离医院的过程中，有哪些经验和成果可以和大家分享，值得我们继续思考。希望通过中南院在抗疫一线奋战过的建筑师、结构工程师和机电工程师的总结，能为有需要的国家抗击疫情带来一些有益的思考和启发。

通过编写本书，仅分享我们的些许经验，如能给相关国家抗击疫情带来一些有益的帮助，我们不胜荣幸。

除权威发布的技术标准外，书中观点为经验总结，非官方权威观点。本书在编写过程中，由于时间和人手有限，难免存在疏漏之处，敬请读者谅解。

全球化时代，病毒是全人类共同面临的威胁。大家携起手来，必将战胜这场疫情。

目 录

附录 医疗建筑精选

I

技术基础篇
TECHNOLOGICAL BASICS

1

雷神山医院

1.1 数字化技术在建筑行业中的应用概述

雷驰荆楚、术济苍生。从2020年1月24日中南建筑设计院接到雷神山医院设计任务，到2月8日雷神山医院正式交付使用，在15天内，雷神山医院以雷霆万钧的速度迅速建成，在遏制新冠疫情蔓延中起到了重要作用[1]。数字化技术精细高效的特点在雷神山医院甚至是建筑设计领域中也起到了明显效果。

1.1.1 建筑业数字化变革的必要性

当今世界，科技革命和产业变革日新月异，数字经济蓬勃发展，深刻改变着人类生产生活方式，对各国经济社会发展、全球治理体系、人类文明进程影响深远[2]。中国制造、中国创造、中国建造共同发力，继续改变着中国的面貌[3]，数字经济支持下的数字化技术与建筑设计行业不断融合，持续产生着"化学反应"，深刻改变着建筑业的生产方式，为建筑业向高质量发展的转型之路奠定了良好的基础。

数字化（digitalization）技术是基于信息化（informa-tization）技术所提供的支持和能力，让业务和技术产生交互，改变传统的商业运作模式，创建新的业务范围，并提供创造收入和价值的新机会。

数字化技术的概念比较宽泛，在建筑业，它以BIM技术为核心，集合了参数化、虚拟仿真、三维可视化、物联网、云计算、移动互联网、大数据、人工智能等各种信息技术，借助各种软件，共同完成建筑模型的数字化交付及存储，实现建筑产品数字虚体和建筑实体的"数字孪生"，高效迅速地完成设计、模拟和优化，得出更好的方案，全面指导建筑设计的开展和进行，以提升效率，降低成本。

在全球进入数字经济时代的背景下，面对我国建筑业高污染、高能耗、低效率的普遍现状，数字化变革已不是选择，而是唯一出路。利用数字化手

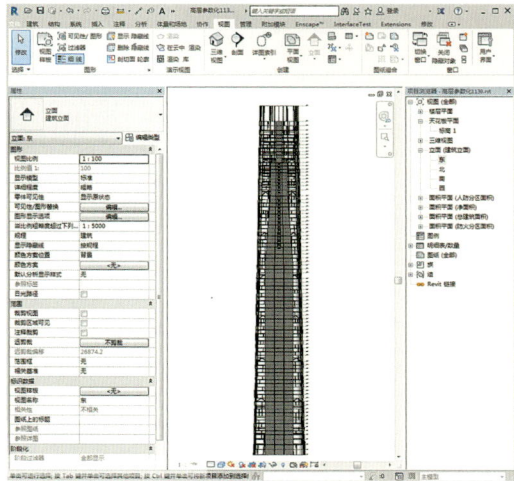

（a） （b）

图 1.1　数字化技术从先进制造业向建筑行业转移
（a）大型先进制造业；（b）建筑行业数字化

段，可在企业层面的信息化系统基础上，建立基于项目的生产作业数字化系统，延伸数字化技术的应用范围，从方案创作直至建筑运维，提高数字化技术的应用成效。在数字化生成、数字化技术分析优化、数字化技术管理等方面深入研发，建立生态，完成统一业务平台"最后一公里"的需求匹配研发，实现建筑企业在生产维度、组织维度和价值链维度的数字化转型升级。

1.1.2　建筑业数字化应用特点

在计算机发明的较早时期，就有部分从业者将建筑结构模型数字化后，通过计算机程序完成复杂的力学分析计算，但数字化技术在建筑业的快速发展和普及则是在以航空工业为代表的大型先进制造业实现数字化以后产生的（图1.1）。究其原因，一方面，早期的数字化建设投资巨大，只有航空工业才能承受这个成本；另一方面，与制造业相比，建筑行业的数字化模型具有数据量大、尺寸跨度大等特点，需要数字化软件、硬件技术发展到一定阶段才能承载建筑业的需求。

同时建筑业和先进制造业也有很多类似的

特征。以飞机制造为例，飞机制造数据量大、变更多、任务种类多、制造成本控制严格、供应链长、供应商多，等等。而对于复杂的建筑工程项目，同样存在信息量大、设计施工修改多、参与专业多、施工成本控制严格、工程周期长、大量的分包商协作等类似特点，使得从大型制造业积累的数字化经验，可以向建筑工程行业进行技术转移。

对于数字化技术在建筑业的应用可以从以下三个方面阐述：首先，数字化技术建立建筑及周边环境的几何模型，存储建筑生命周期信息，建立建筑全生命周期的业务流程并以更高效的方式完成建筑的数字化表达。其次，数字化技术可以借助高效的计算机系统实现建筑行业的高效生成、精准分析和高质量管理。最后，数字化技术可以通过技术革新重塑建筑行业的商业模式，为建筑行业未来发展催化赋能。

1.1.3　建筑行业的数字化表达

建筑物的数字化表达，首先是几何外形和结构组成的数字化和可视化。从手绘图纸到二维CAD绘图再到逐步普及的BIM模型（图1.2），

图 1.2　建筑三维 BIM 模型

图 1.3　虚拟现实

来源：https://pixabay.com/zh/illustrations/vr-virtual-reality-vr-goggles-6770800/

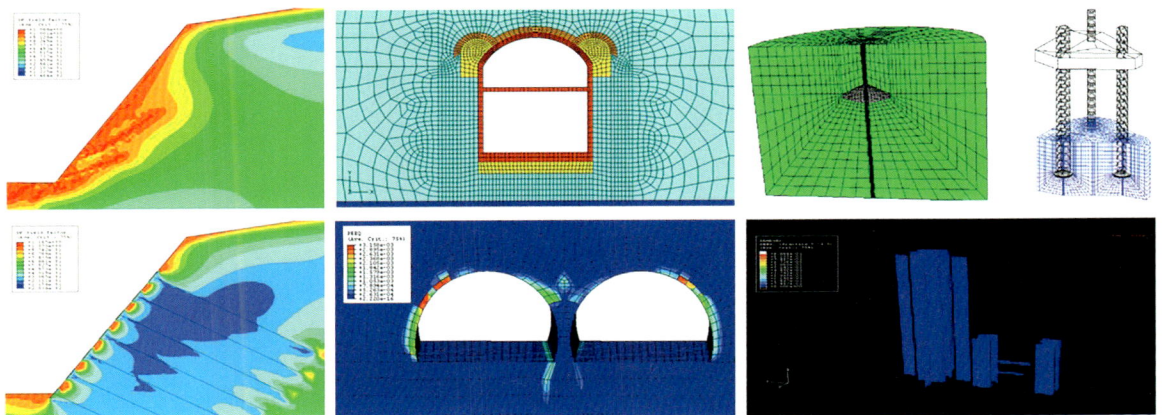

图 1.4　计算机辅助工程（Computer Aided Engineering，CAE）分析结果数字化表达

乃至虚拟现实，建筑的数字化表达越来越符合人类的观察习惯。

　　而虚拟现实、增强现实技术则在体验上更进一步，它们通过计算机在虚拟空间中模拟出数字化建筑，为使用者提供逼真的视觉感知（图 1.3）。使用者作为主角沉浸在模拟环境中，获得与真实世界相同或相似的感知，产生强烈的临场感。虚拟现实、增强现实技术在提供真实空间感受的同时，还可以借助科技的力量，展现光照、热量、能源、风力、电场等现实世界中不可见的数字信息，协助使用者提高认知，深化概念，萌发创意，启发创造性思维。

　　建筑三维数字化实现后，工程计量自然水到渠成，通过提取建筑三维数字模型中的信息，我们可以获取建筑物的任意阶段、时间点、系统、楼层、专业等属性并提取使用，无论哪一层、哪一段，竖向构件还是水平构件，何种规格型号，均可快速提取。

　　建筑行业涉及专业众多，除建筑专业外，还有结构、给水排水、暖通、电气，等等，而对应的知识学科则包含物理（力学、热学、电磁学、光学、声学等）、化学、生物、地理、地球物理等。在建筑三维模型的基础上，通过数字化技术能实现多专业、多学科的数字化知识表达（图 1.4）。

　　工程人员可以从数字化建筑模型中提取自己专业所需信息，进行各种分析、模拟和优化，并根据仿真计算结果对项目设计方案进行调整，之后再对新方案进行计算，直到满意的设计方

图 1.5 建筑全生命周期管理平台

图 1.6 BIM 中的 5D 模拟

案产生为止。典型的应用场景有：高层建筑结构的振型分析、高层剪力墙弹塑性动力分析、框架结构地震响应分析、隧道开挖、施工过程对环境的影响、边坡稳定的剪切带计算，等等。

数字技术向建筑行业下游阶段延伸，进入施工甚至运营阶段，可以为整个建筑工程项目的全生命周期创造价值。建筑全生命周期管理即 BLM（Building Lifecycle Management），覆盖建筑工程项目从规划、设计到施工，再到运营维护，直至报废拆除为止的全生命周期过程（图 1.5）。建筑工程项目具有投资总额大、技术含量高、施工周期长、风险程度高、涉及单位众多等特点，因此建筑全生命周期管理就显得十分重要。

建筑全生命周期管理的理念让建筑项目相关各方都要具备全局观念，在设计阶段就要通过施工模拟来考虑施工阶段的可建造性，让建筑物具备更好的运营和维护的方便性和经济性。

施工仿真技术可以对工程施工方案进行虚拟验证和优化，可以实现施工工艺验证、施工进度计划验证、关键工艺验证、运动干涉检查等，避免错、漏、碰、缺，排除施工过程中的各类冲突及风险，对比分析不同施工方案的可行性，使建筑工程施工成本得到有效控制，极大程度地降低施工安全事故的发生概率。

施工仿真还可以在由三维空间和时间构建的 4D 模型的基础上增加成本的维度来建立 5D 模型（图 1.6），通过 5D 模型来实现精细化的预算和项目成本可视化。通过对工程项目进行 5D

仿真模拟，可以得到所有建筑构件的准确工程量，实现工时、成本的统计分析，进而实现建筑工程造价精准控制。

在建筑物建成以后，通过建筑全生命周期管理平台实现施工方向业主或运营方的完整数字化交付。建筑物进入运营维护阶段。

在运营维护阶段，在建筑全生命周期管理平台上实现维护维修作业指导书、维护维修任务分配、设备参数信息查询等。与物联网技术结合，各类设备的状态信息、管路压力、环境的 $PM_{2.5}$ 数据、温度、湿度、能源消耗等数据通过传感器的部署和采集，在管理平台上以可视化的方式呈现，如图 1.7 所示。

数字化技术除了支持建筑物本身以外，还可以将建筑物周边的道路、相邻建筑等数字化，形成建筑及周边环境的区域级数字化模型。当整个城市的建筑都实现数字化时，数字化模型就从单个 BIM 模型扩展为 CIM（City Information Modeling），即城市信息模型。

借助于数字化技术，可以在虚拟空间构建一个与物理世界相匹配的孪生城市，以数字城市模型为基础，对城市运营和治理进行科学决策。以虚拟新加坡（图 1.8）为例，隶属于总理办公室的新加坡国立研究基金会（National Research Foundation，NRF）基于地形学数据和实时动态数据为新加坡建立三维数字孪生城市，供城市规划人员在虚拟环境中模拟对创新解决方案的测试。除常规的地图数据和地形数据外，该孪生城

数字化应急医院设计及建造技术

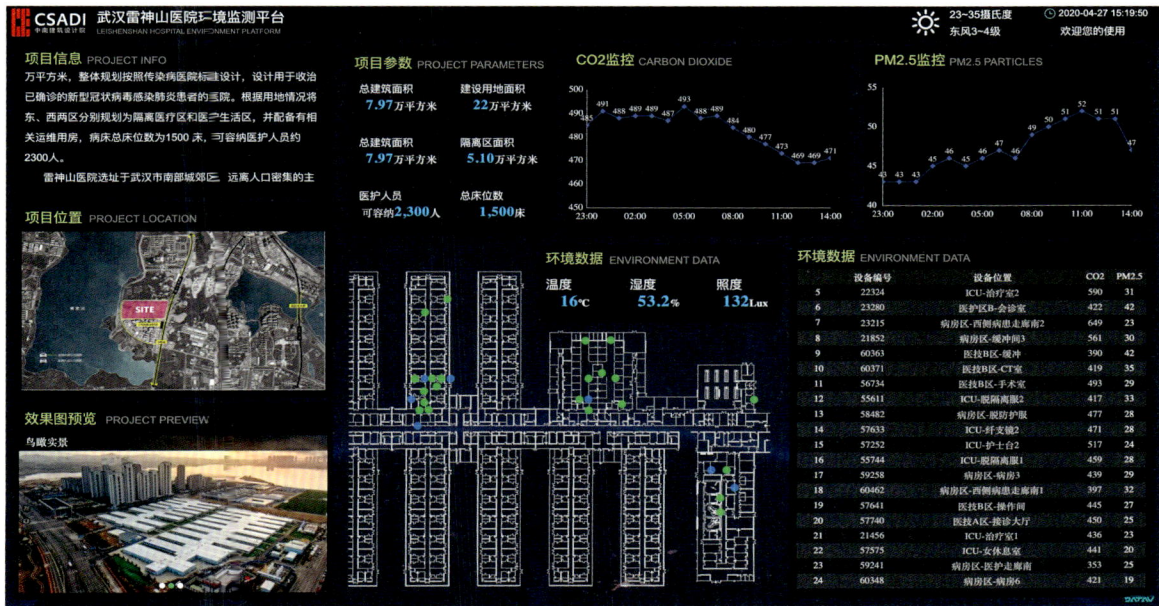

图 1.7　建筑运营维护管理

图 1.8　数字化城市：虚拟新加坡

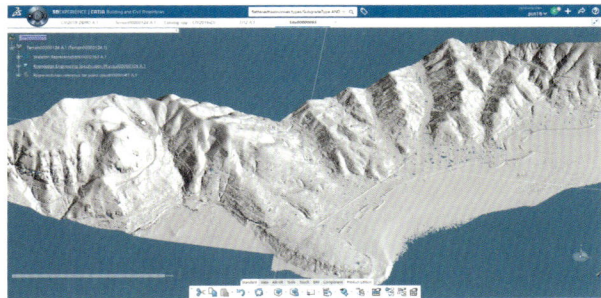

图 1.9　数字地形模型

市还整合实时交通、人口、移动通信、学校、卫生、能源、房产和气候数据等，便于城市规划人员展开虚拟实验。如结合流体力学仿真可以模拟建筑物周边、街道和绿地的气流，建筑设计时为残障人士和老龄人口规划无障碍路径等。

通过导入测量点、激光雷达（Light Detection and Ranging，LiDAR）、点云、倾斜摄影等数据，生成地形网格面，数字化技术可以生成数字地形模型（图 1.9），建立真实的地理环境，还原真实的建筑场地，进而完成自动生成等高线数据、高程着色、流域及地表径流模拟、场地布置及土方规划等工作。

1.1.4　建筑行业数字化技术提质增效

在实现建筑行业数字化表达的基础上，数字化技术将实现建筑业技术发展的催化赋能，涉及利用数字化技术实现高效生成、利用数字化技术实现精准分析、利用数字化技术实现高质量管理等方面。

1. 利用数字化技术实现高效生成

通过数字化技术支持的参数化设计工具，可以在 BIM 平台按照指定的逻辑关系，自动完成装配式建筑部品的选择组合、建筑曲面造型、网格创建等复杂操作，生成完整精确的数字化设计模型。

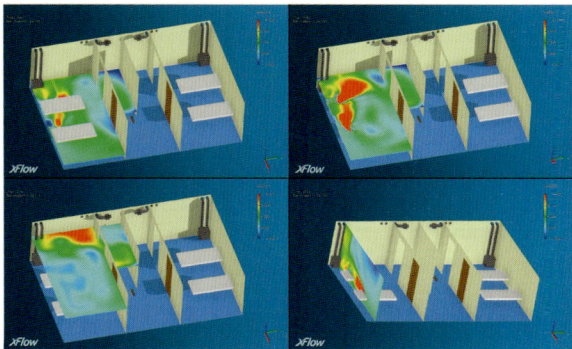

图 1.10　建筑物内污染物扩散模拟

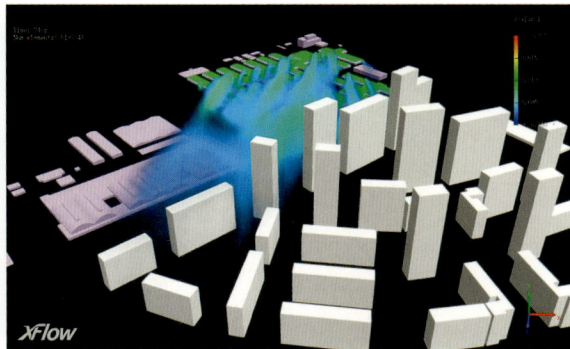

图 1.11　场地污染物扩散模拟

借助参数化设计技术，数字化技术设计师可以将影响建筑设计的要素信息定义为各种参数，并将各个参数组织串联，形成完整的设计系统，在设计过程中，对函数进行分析和优化，并采取适合的算法，获得不同的建筑设计方案，并最终选出较优的设计方案。

参数化设计方法具有逻辑性强、整体化程度高的特点，相比于一般设计方法，参数化设计形成的模型具有定义明确、逻辑清晰、模型精准、形态多样、一致性和整体化程度高的特点。通过参数化技术，设计师可以快速构建高精度数字化建筑模型，并在模型关键参数间建立联动关系，对建筑方案进行比选和优化重构，全面提升工程的设计水平和设计质量。

2. 利用数字化技术实现精准分析

数字化技术不仅在模型生成上表现卓越，而且在项目优化中也具有传统手段无可比拟的优势。

数字化技术可以集成利用 BIM、物联网、机器学习的最新成果，借助计算机系统的高速运算能力，构建、优化目标函数，用数值分析代替物理实验，不断修改迭代从而改进系统的运行效能，为建筑业全产业链提供优化升级的技术支撑。

借助 CFD 技术，可以模拟建筑物内的气流组织和污染物扩散情况（图 1.10），优化通风系统布置方案；可以建立城市级的大尺度流场模型，模拟周边风环境和污染物扩散的影响（图 1.11），为场地规划和建筑布局提供设计依据。

大型公共场所或者重要公共设施等人群集中的场所，设施复杂、人群庞大，如何制定有效的应急疏散预案，在高峰期和紧急情况下科学、有效地疏导人群逐渐成为一个复杂的问题。传统人群疏散预案的制定对专家经验的依赖度较高，而数字化技术则可以通过建立数值模型，模拟人群疏散的过程，展示人群疏散场景（图 1.12），在建筑设计阶段优化空间布局，在建筑使用阶段制定相应的疏散预案，从而更加全面、准确、细致地进行分析、研究。数字化技术支持下的人群疏散模拟具有形式直观、经济高效、人员风险小、反馈快速及时等优势。

通过数字化技术，我们可以在建筑的 BIM 模型中计算、模拟太阳运行轨迹，实时计算室内室外日照时长分布情况（图 1.13），优化建筑排布，以更好地满足规范要求。

数字化技术的广泛应用，为跨专业协同提供了有利的技术条件，通过 IFC（Industry Foundation Classes，建筑国际化工业标准）、COBIE（Construction Operations Building Information Exchange，施工运营建筑信息交换）等标准，数字化技术可以在专业软件间建立数据接口，构件专业模型的数据联动机制，形

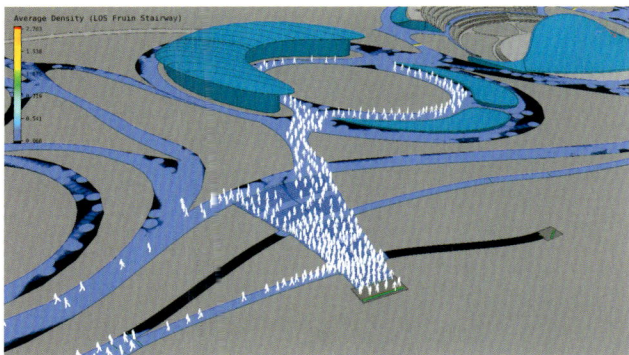

图 1.12　人群疏散分析

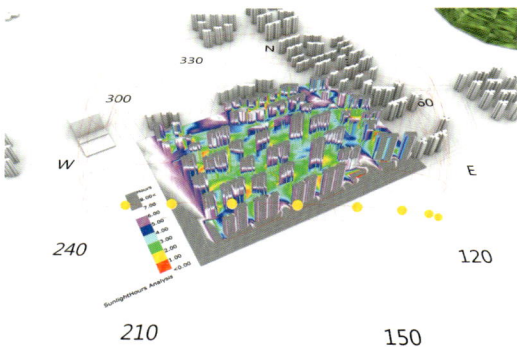

图 1.13　场地日照分析

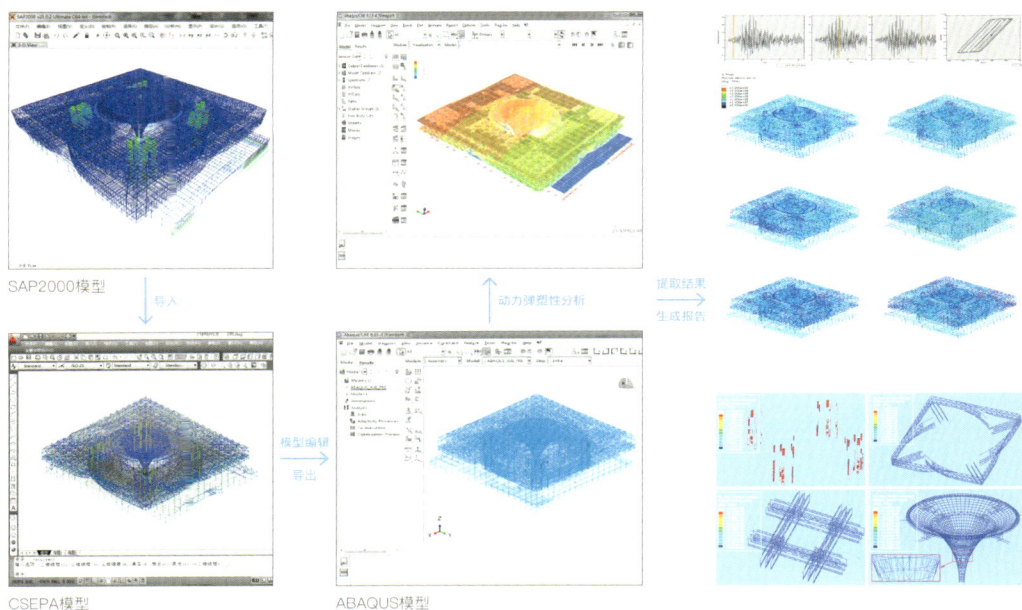

SAP2000模型

CSEPA模型

ABAQUS模型

动力弹塑性分析

提取结果
生成报告

导入

模型编辑
导出

图 1.14　模型转换

成多专业一体化设计平台（图 1.14），为建筑方案的快速分析、结构优化等工作提供更快的方法，创造更好的条件。

3. 利用数字化技术实现高质量管理

利用数字化技术，可在工程设计阶段快速建立多专业三维模型，搭建协同设计平台，进行设计管理。数字化技术可以在工程设计中快速检查各种错、漏、碰、缺，可随时在三维场景中观察、漫游，发现各种设计问题并给出解决方案，这样就可以大幅减少施工过程中的返工次数，避免了资源、时间、成本的浪费。

数字化技术能够汇集建筑工程项目全过程中的各类信息如三维模型、图纸、合同、文档等，形成单一数据源，连接项目相关的所有主体，包括业主和开发商、总承包方、设计院、施工方、物业管理、建材厂商、设备供应商等，为用户建立跨专业、跨企业的协同工作环境，为不同用户提供三维仿真设计、分析、制造和建造的能力，实现全流程的项目管理、产业链的沟通协作和数据的高效流转，对工期、质量、造价进行严格精准把控（图 1.15）。

在项目实施现场，将施工业务与数字化相

图 1.15　基于单一数据源的建筑全生命周期的高质量管理

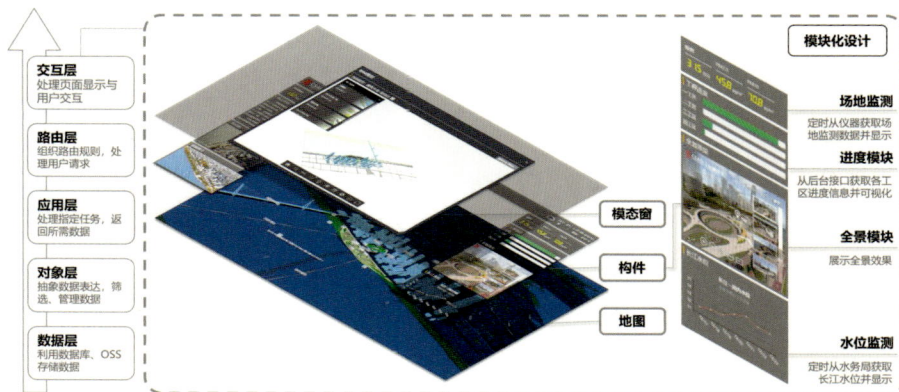

图 1.16　数字化技术支持项目施工管理

结合，利用图形技术、识别技术、视频监控技术、移动互联技术、物联网技术等数字化技术实时采集现场信息，自动完成进度跟踪、材料管理、人工排班、关键节点预警、资金管理等工作，实现综合研判、高效管理（图 1.16）。

1.1.5　数字化技术赋能未来

随着数字化技术与建筑行业的高度融合，建筑行业的开发流程、管理模式、生产方式、供应链等均发生了深刻改变。数字化技术将推进建造方式的创新，更新、改造和升级建筑业产业链，重塑建筑行业的商业模式。

首先，数字化技术将赋能建筑行业生产力发展。建筑行业装配率逐年提升，建筑行业标准化、模块化、装配化的趋势明显加强。工厂智能化生产、自动化配送将广泛普及，整个建筑行业手工劳动替代率逐步提升，准确率和效率不断提高，建筑产品质量因生产线的加工模式得以保证，质量通病基本消除，大量危险作业被人工智能机器替代，安全事故等风险基本消除。

精确定位的三维模型直接与自动化生产线对接，各项用料信息，如产品编号、材质、数量、重量、价格等也可直接由模型提取，通过多种方式将数据输送至生产线。

随着自修复混凝土、气凝胶和纳米材料等新型建筑材料的应用，基于建筑三维模型的 3D 打印和预组装等创新的施工方法将会逐步应用，进而降低成本，加快施工速度，同时提高质量和安全。

其次，数字化技术赋能建筑智能感知，持续

数字化应急医院设计及建造技术

优化。作为数字化技术与建筑艺术相结合的产物，智能建筑将逕通过传感器和物联网完成数据联通和信息采集，从而具备持续、全面的感知能力，再经过人工智能"大脑"思考决策，并通过反馈不断自我优化，迭代更新，为智慧建筑赋予生命力，使之能够随着组织和人的发展变化而变化，与人互动，时刻感知、预判，并主动响应人的个性化需求。

最后，数字化技术赋能建筑行业，形成完整的数字生态圈。建筑产业链上下游均由专业化公司构成，例如预制构件工厂、物资配送中心、机电设备公司、专业共应商等。通过建设建筑行业工业互联网，接入建筑产业链各类公司，建筑工程总承包方和业主一起履行投资、设计、施工、运营、维护的完整生命周期职责，通过建筑行业互联网找到合适的下游供应商，建筑产业链下游企业均通过建筑生命周期管理平台实现互联互通、资源共享，共同打造出完整的业务体系、管理体系，智慧化的运营和管理体系得到高效运转，生产效率和效益得到极大的提升，建筑物的绿色节能、低碳环保均达到空前高度。

1.2 科学选址，快速建造

1.2.1 确保环境安全、生物安全的科学选址

2020年1月下旬，新冠肺炎患者确诊人数激增。面对严峻的疫情，为提高新冠肺炎患者的收治能力，打赢武汉保卫战，1月25日下午，武汉市决定半个月之内在长江以北的火神山医院之外、长江以南再建一座规模更大的雷神山医院。项目总建筑面积经过三次调整增加，最后规模确定为：总建筑面积约8万平方米，病床总床位数约1500张，可容纳医护人员约2300人。2月8日晚8点，雷神山医院正式建成移交，并迎来首批新冠肺炎患者。

雷神山医院选址于江夏区军运村，三环线与四环线之间（图1.17），除满足国家相关规范对应急医院建设选址的要求外，项目选址有如下几方面的特殊优势：

（1）远离人群密集的中心城区。武汉市中心城区人口密集，用地紧张，基地远离人群密集区，能够有效地降低感染更多人群的风险，同时也避免了对中心城区的环境污染。

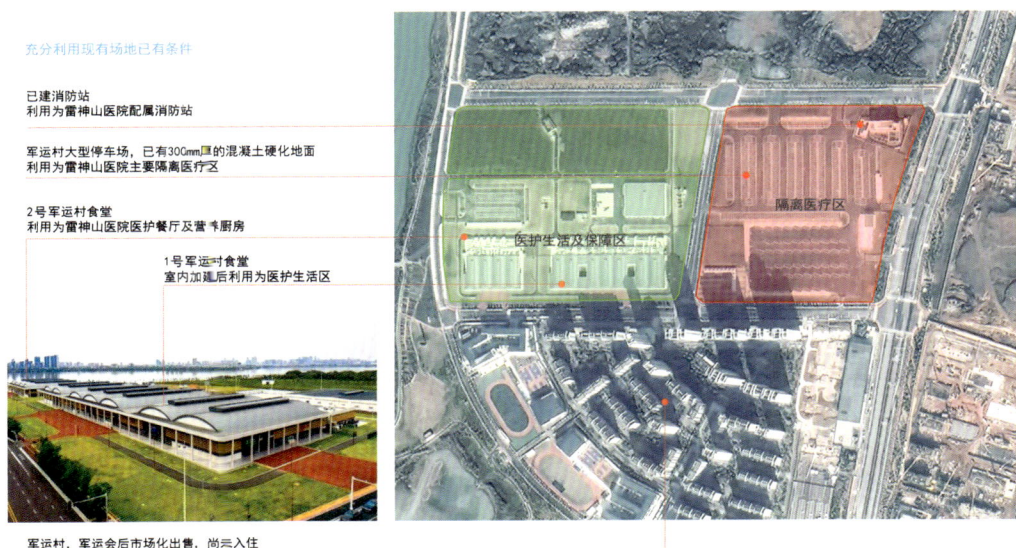

图1.17 雷神山医院选址情况

来源：军运会 万人食堂曝光，每天将消耗食材数十吨 [N/OL]. 腾讯网，2009–10–28[2021–12–15]. https://new.qq.com/omn/20.91028/20191028A0DM8700.html?pc（左下图）；基于百度地图自绘（右图）

（2）基地处于城市下风向。武汉市冬季主导风向为北风和东北风，选址的军运村地处武汉市中心区域以南，属于城市冬季下风向位置。

（3）周边地块基本处于开发建设中，利于在空间上进行卫生隔离，能最大限度地减少对周边环境的影响。场地南面为原军运村运动员宿舍，后期考虑商业出售，尚无人员入住；东面为城市待开发用地；北面为城市绿地；西面为闲置的军运村后勤区，包括可供万人使用的运动员餐厅、工作人员餐厅、附属物资仓库。三栋建筑主体均为1层大跨度建筑，内部空间较大，便于建筑内部灵活改造。

（4）场地交通便利。东邻主干道黄家湖大道，距离三环、四环线均约3km，距武汉三镇仅30min车程；黄家湖大道有双向八车道，利于整个区块的车辆进出疏散。场地四周交通设施完善，道路环通。现有的军运路、兴军路、强军路具备在医院运营期间可封闭管控的优势，而军体路则可以被征用作为医院院区内部道路。场地内外良好的交通条件和离中心城区较近的交通优势，能够满足疫情期间大量患者转诊和大批物资、设备快速运送等需求。

（5）基础设施完备。2019年10月军运会结束后闲置下来的项目用地是一块市政配套设施完善的场地，水、电、气、通信、5G网络等市政基础设施完备。设计用量能够满足雷神山医院运营量需求，同时能满足能源供应、远程智控、数据传输、高效医疗及生活保障的需求。

（6）具备快速建设场地的条件。选作隔离医疗区的场地东区为大型停车场，已有300mm厚的混凝土硬化地面，易于集装箱模块化装配

式建造，并可作为隔离医疗区的基础整浇层，大大减少场地平整的工作量，为雷神山医院的建设争取了时间。

（7）消防设施完善。用地东北角为基础配套保留的消防站，可作为消防应急设施使用，使项目在消防安全上得到充分保障。

作为一座收治新冠肺炎患者的应急医院，选址首先考虑了环境安全与应急交通需要。同时，合适的场地选择也可减少项目工程量，特别是早期的地基处理工程量，为项目的建设奠定了良好的基础[4]。

1.2.2 模块化设计

雷神山医院总体规模相当于两个火神山医院，但工期却与火神山医院相当，建设工期异常紧张，医院得以快速建设离不开如下三个重要因素，即模块化设计、标准化生产、装配式建造，如图1.18所示。

模块化的设计理念贯穿雷神山医院设计的始终，可以归纳为三个主要层面：宏观的规划架构、中观的功能布局、微观的房间设计。

1. 模块化的规划架构

雷神山医院项目整体分为隔离医疗区、医护生活区和保障功能区三大模块，其中隔离医疗区为主体功能区，如图1.19所示。

隔离医疗区采用鱼骨状布局，根据地形分为南北两区，每个区域均以中央的工作人员通道作为中央主轴，各病房单元、医技单元作为功能模块拼接到中央主轴上。北区设15个病房模块，南区设15个病房模块，模块间距12m，其他功能模块分别为：30张病床的ICU模块（南北区各1个）、检验科模块、手术室与CT检查室模块、药房药库模块、集中卫生通过模块（南北区各1个）。功能模块总数为北区20个，南区17个，如图1.20所示。其中，病房模块为隔离医疗区主体功能模块，每个病房模块设50张床位，各

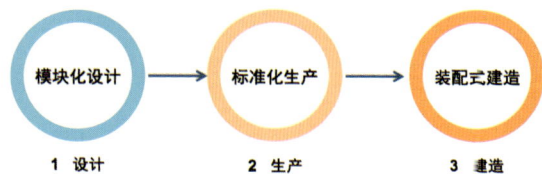
图1.18　环环相扣：雷神山医院建设的三个阶段

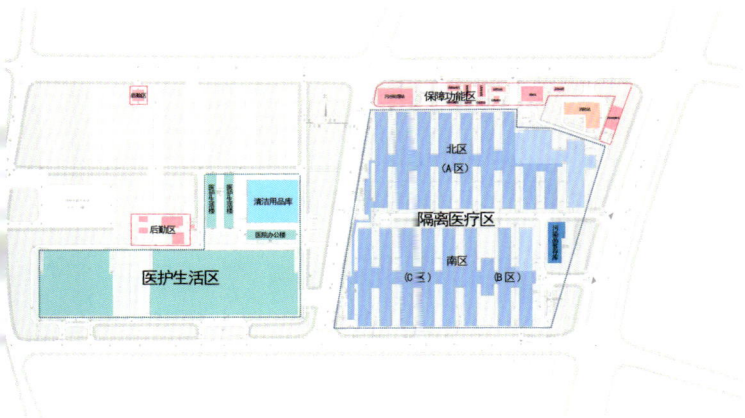

图 1.19　项目整体功能分区图

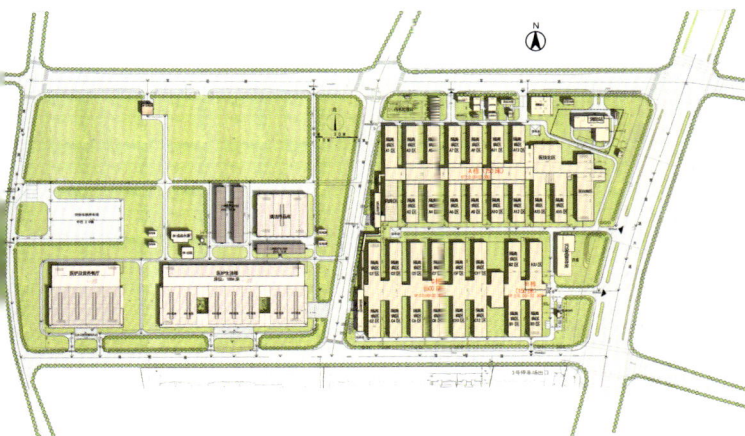

图 1.20　项目总平面：隔离医疗区鱼骨状布局模式

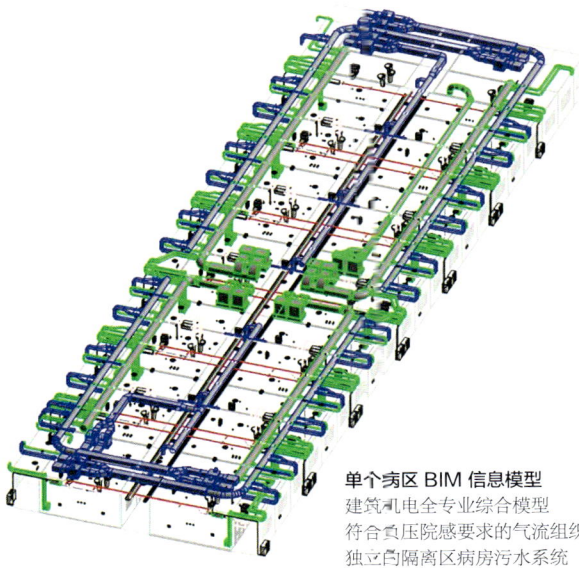

单个病区 BIM 信息模型
建筑机电全专业综合模型
符合负压院感要求的气流组织和暖通设计
独立的隔离区病房污水系统

图 1.21　病区模块及其医疗流程

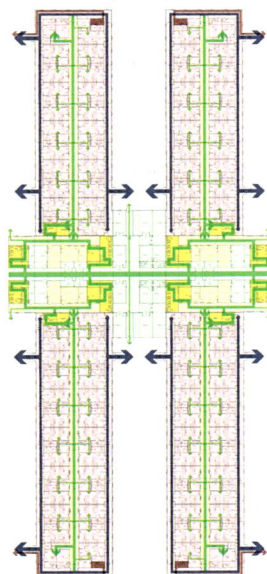

图例说明：
患者流线
医护流线
污物流线
洁净通过
污物间
清洁区
半污染区
污染区

病区组合后的医疗流程
鱼骨状布局模式
"三区两通道"建筑布局

病房模块之间具有通用性和一致性，遵从"三区两通道"与气流组织设计原则，如图 1.21 所示。

各模块边界清晰，功能独立，通过中央主轴相互联系，便于齐头并进分别设计、分别建造，节省设计与施工时间。

2. 模块化的功能布局

中观层面，医疗功能流程通过模块化的功能布局得以快速实现。为减少功能模块类型，降低施工差异，提高施工效率，雷神山医院医疗区 95% 的区域主要由三种基本功能模块组成，分别如下。

（1）由 2 间病房及其合用的缓冲前室组成的病房基本模块，即污染区主要功能模块。

（2）由男女卫生通过、护士站、配药室、医生办公室、传递间等组成的各病区医护工作区，即半污染区主要功能模块。

（3）由值班室、医护卫生间、药品库、仪器库、耗材库、配电间等组成的医护辅助区，即清洁区主要功能模块。

三种基本功能模块内部统一标准，相互拼合成 30 个病房护理区。这三种基本功能模块也分

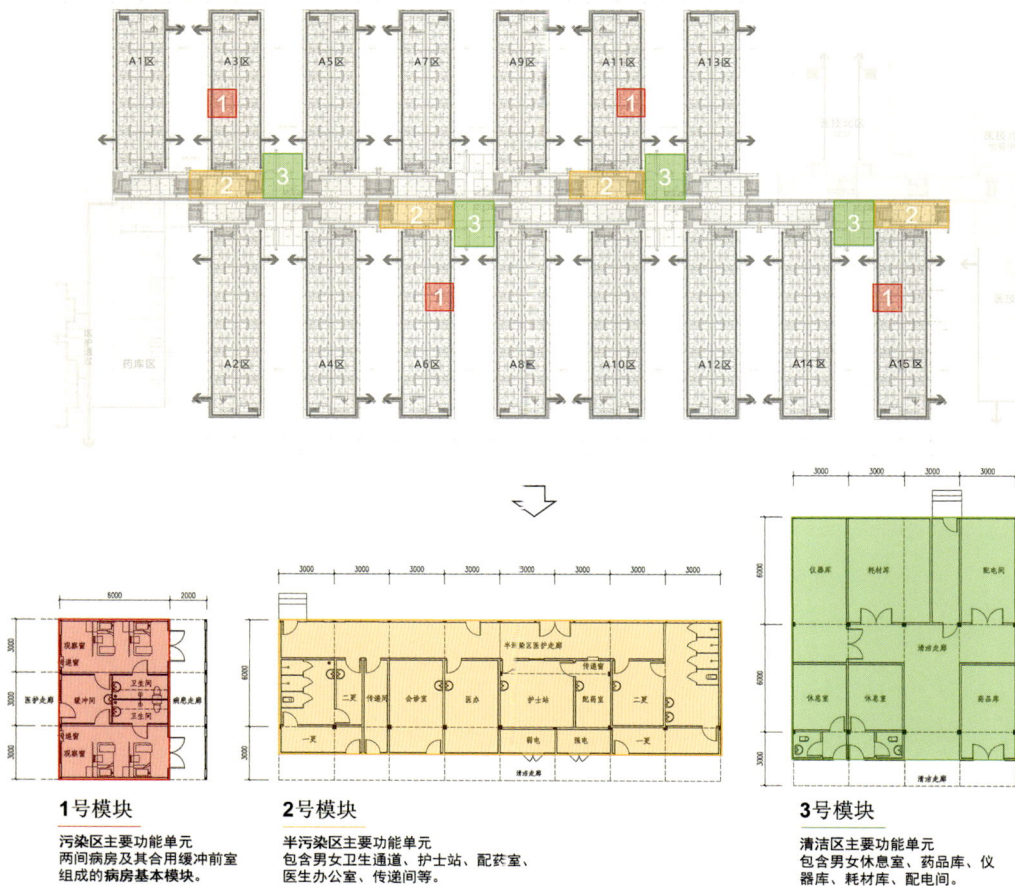

1号模块
污染区主要功能单元
两间病房及其合用缓冲前室
组成的病房基本模块。

2号模块
半污染区主要功能单元
包含男女卫生通道、护士站、配药室、
医生办公室、传递间等。

3号模块
清洁区主要功能单元
包含男女休息室、药品库、仪
器库、耗材库、配电间。

图1.22 三种基本功能单元

别对应了医疗流程中重要的三个基本分区：污染区、半污染区、清洁区。对三种基本功能模块的精细化设计，使每个功能模块的医疗流程、建造方法、机电管线既满足应急医院的高标准流程要求，也能做到施工建造、机电安装简便快捷。只有做好了最底层的基础模块，才能保证规模巨大的雷神山医院在快速的建造过程中始终保持较高的整体品质，满足应急医院医疗流程高标准要求，如图1.22所示。

3. 模块化的房间设计

落实到每个房间时，为便于标准化生产，通过模块化划分，把整个隔离医疗区（含药房药库及集中卫生通过等）分解为A、B两种集装箱式板房骨架模块，其中，A模块尺寸为3m（宽）×6m（长）×2.9m（高），总数量为1918个（北区

970个，南区948个）；B模块尺寸为2m（宽）×6m（长）×2.9m（高），总数量为990个（北区495个，南区495个），如图1.23、图1.24所示。所有模块由工厂预制生产，运输到场地周边道路进行场外组装，现场吊装到位，病房模块内部小型管线同步预留预埋。部分特殊功能房间（如药房药库）以箱体为基本单元，通过水平及竖直方向的不同组合形成宽敞的使用空间。

同时，对主要功能房间进行重点模块化设计，特别是与应急医院医疗使用功能和医疗流程密切相关的房间，主要有如下几类：①病房基本模块，如图1.25所示；②配电间、小型配套卫生间等设备、设施用房，如图1.26所示；③连接不同分区的卫生通过等关键院感控制用房，如图1.27所示。

■ 尺寸：3m(宽)X6m(长)X2.9m(高)　数量：北区970个　　■ 尺寸：2m(宽)X6m(长)X2.9m(高)　数量：北区495个

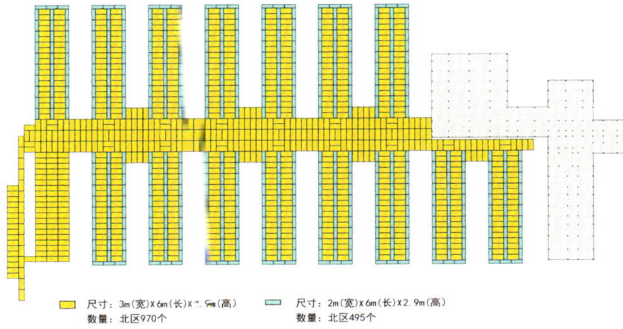

图 1.23　北区两种集装箱式板房模块尺寸与分布图

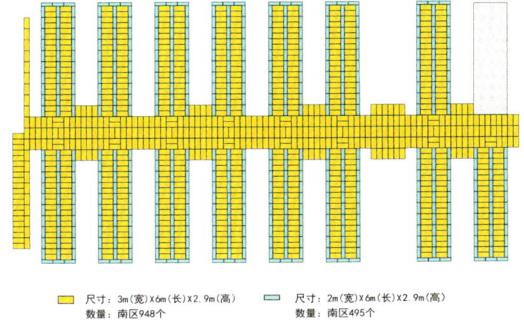

■ 尺寸：3m(宽)X6m(长)X2.9m(高)　数量：南区948个　　■ 尺寸：2m(宽)X6m(长)X2.9m(高)　数量：南区495个

图 1.24　南区两种集装箱式板房模块尺寸与分布图

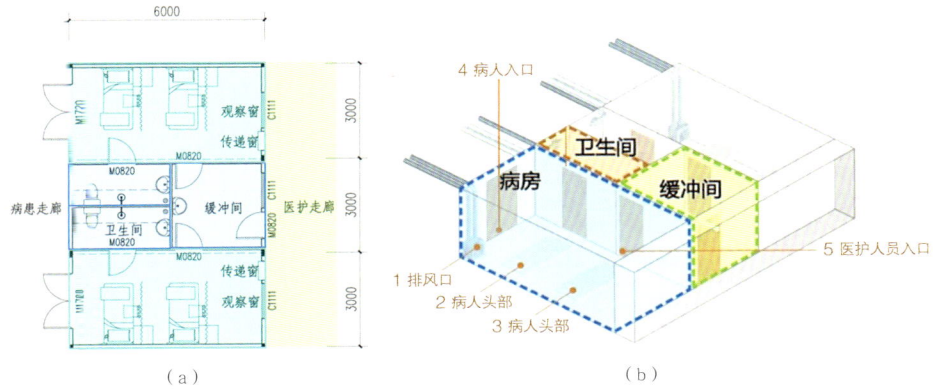

（a）

4 病人入口
卫生间
病房　　缓冲间
5 医护人员入口
1 排风口
2 病人头部
3 病人头部

（b）

图 1.25　病房基本模块：满足医患分流、缓冲通道、气流组织需求
（a）平面图；（b）示意图

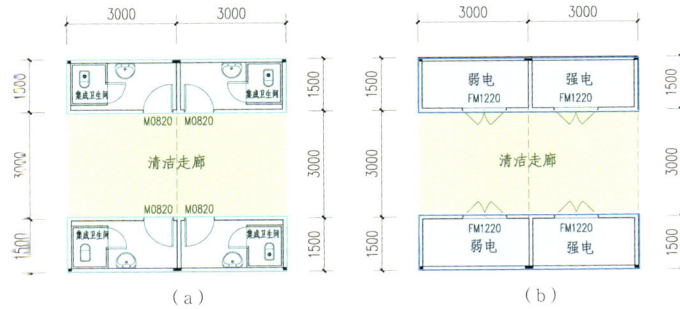

（a）

（b）

图 1.26　配电间、小型配套卫生间等设备设施用房
（a）医护小型卫生间，南北两区共7组；（b）配电间，南北两区共16组

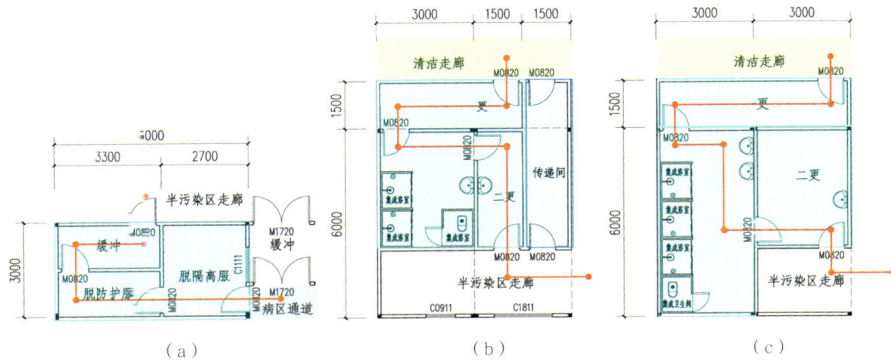

（a）

（b）

（c）

图 1.27　连接不同分区的卫生通过等关键院感控制用房
（a）脱衣缓冲，南北两区共30组；（b）男卫生通过，南北两区共32组；（c）女卫生通过，南北两区共32组

图 1.28　江苏省集装箱式板房生产厂家援建生产雷神山医院集装箱式板房

来源：武汉雷神山医院首批病房从苏州启程 [N/OL]. 中国新闻网，2020-01-27[2021-12-15].
https://www.chinanews.com.cn/sh/2020/01-27/9070773.shtml.

图 1.29　安徽的钢结构厂家仅用 40h 就将 2000 件标准化钢结构构件装车发往雷神山医院工地

来源：安徽 40 小时赶制 2000 件钢构件 援建雷神山医院中安在线 [N/OL]. 中安在线，2020-02-16[2021-12-15].
https://baijiahao.baidu.com/s?id=1658697738401003020&wfr=spider&for=pc

1.2.3　标准化生产

在模块化设计的基础上，在现场处理地坪和预埋管网的同时，各部件的标准化生产已在全国多家工厂同时展开。现场除地基处理、主干管网预埋与 CT 室混凝土外墙工程外，几乎都采用干作业法施工。雷神山医院 95% 以上的建筑用材均为工厂生产的预制构件，如隔离医疗区板房支墩，钢结构箱式模块化房屋（即集装箱式板房，如图 1.28 所示），门窗，钢结构构件（图 1.29），成品卫生间与淋浴间，医技区内墙隔断，手术室、检验室无菌墙板，所有卫浴产品和机电设施。

基于统一的技术规范，标准化、工业化的生产，工厂的环境及加工精度，各部件质量得到控制，缺陷减少，在现场的快速组装中能确保整体施工质量，现场减少湿作业也能使多工种安装同时展开。此外，标准预制构件还可循环利用，可拆卸重新组装，节能环保。

1.2.4　装配式建造

雷神山医院建设最大的挑战就是时间紧、任务重，相当于要把两三年的建设任务在十余天内完成。装配式的建造方式是项目建设的必然选择。项目针对隔离医疗区住院部、隔离医疗区医技部、医护生活区不同的建筑功能和空间高度要求，采用了三种不同的装配式建造方式。

1. 住院部装配式建造方式

隔离医疗区住院部是整个项目中占比最大的功能用房，约占总体工程量的 70%，全部采用较为成熟的钢结构箱式模块化房屋作为基本建造单元，相对于传统彩钢板房，其防火、抗震、

图 1.30　钢结构箱式模块化房屋现场装配

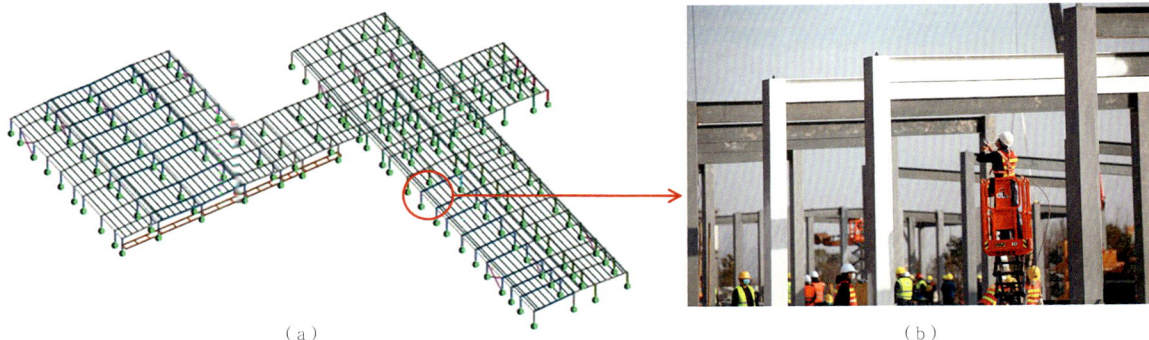

（a）　　　　　　　　　　　　　　　　　　　　（b）

图 1.31　隔离医疗区医技部钢框架结构体系
（a）医技区结构形式示意（Midas 分析模型）；（b）现场组装医技区钢结构预制件

隔声、密封防水、保温性能更好。所有模块构件均在工厂进行标准化制作，房屋构件通过标准化连接件在现场进行快速安装，完成房屋整体的建造，如图 1.30 所示。

2. 医技部装配式建造方式

隔离医疗区医技部主要包括 ICU、CT 室、手术室、检验室等。医技楼作为检查、诊断和治疗的综合性大楼，是雷神山医院的"大脑"，功能用房较为复杂，需要的跨度和层高较大且平面柱网不统一，结构跨度通常需在 7m 以上，局部跨度达到 18m，层高为 4.2m 以上。箱式房、板式房等体系难以满足要求，故选择采用钢框架结构体系。钢结构设计和节点深化设计同时进行，钢结构构件经工厂预制生产后，运输到现场拼接组装，如图 1.31 所示。

3. 医护生活区装配式建造方式

医护生活区为 2 层，根据建筑平面和空间要求采用了不同于病房区的轻钢活动板房体系。建筑平面布置以墙板宽度 1820mm 为模数，标准房间单元平面尺寸为 3640mm×5460mm。轻钢活动板房主体结构为轻型钢框架，框架间设有交叉拉索以保证结构刚度和稳定。活动板房技术体系成熟，采用标准化、模数化设计，安装和拆卸非常方便，作为室内临时建筑具有明显优势，平面布局较为灵活，也能满足建筑的各项功能性需求，如图 1.32 所示。

1.2.5　设计体会

作为"时间就是生命，功能大于一切"的建筑，在短短的十余天时间里，雷神山医院高

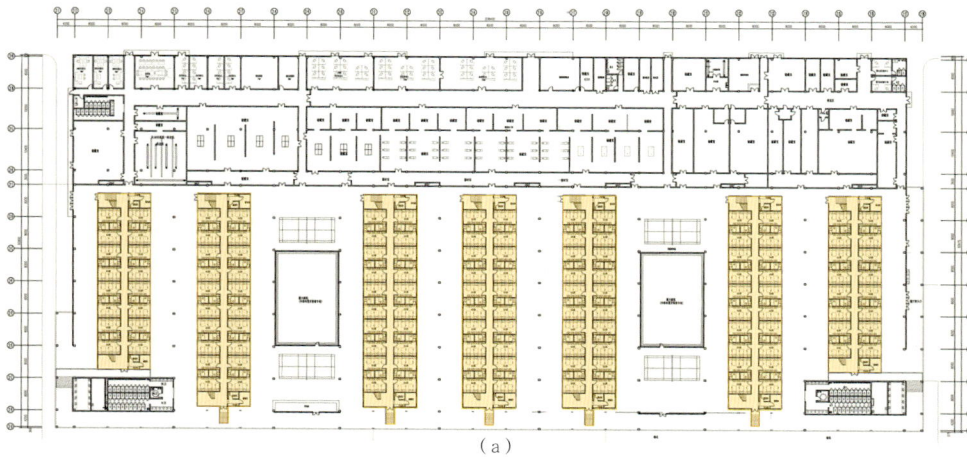

（a）

（b）

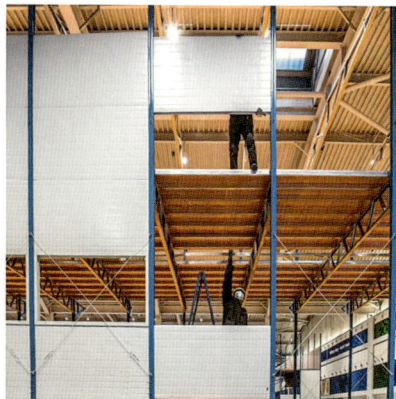

（c）

图 1.32　医护生活区建筑平面布置图、轻钢活动板房区域及建构体系

（a）建筑平面布置图；（b）轻钢活动板房区域；（c）建构体系

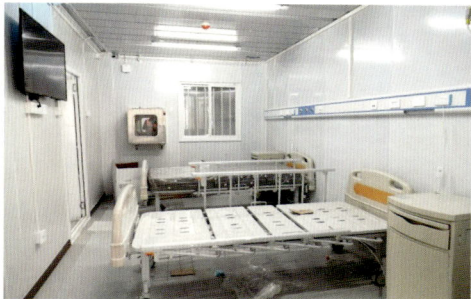

（a）

（b）

（c）

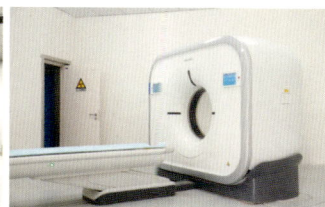

（d）

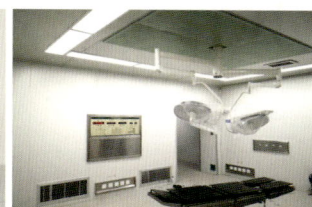

（e）

图 1.33　雷神山医院内部照片

（a）病房；（b）患者走道；（c）检验室；（d）CT 室；（e）手术室

图 1.34　雷神山医院建设实景图

效且高质量建设成为满足收治新冠肺炎患者要求的应急医院（图1.33、图1.34）。前面已从"设计—生产—建造"三个关键阶段阐述了模块化设计、标准化生产、装配式建造的重要作用。

雷神山医院建设也是一次对信息化、数字化设计与装配式建设能力的大考与检阅，在短时间内能高标准建造起来，得益于国家完善的现代化工业，得益于设计单位和施工单位的产业链整合能力，得益于国产医疗配套厂家近些年来的快速发展。

同时，本项目也为应急医院装配式建造的发展提供了一些经验与启示：虽然在雷神山医院建设中，各建筑模块的工业化和装配化已达相当高的水平，但建筑和机电设备之间的集成仍然靠现场安装，工作量较大。医疗设备的集成化水平较低，如病房密闭传递窗的安装、穿墙管洞的密封、病房医疗设备带的现场安装也非常费时。未来可针对负压病房单元、应急医院污水处理单元等关键功能模块进行标准化研

发和生产。可编制此类功能单元工业化生产标准与图集，储备相关生产厂家和产能，在突发疫情时可快速生产标准化功能模块，更为快速地拼建符合收治标准的应急医院，疫情结束后，标准化构件可拆卸组装，循环利用，节能环保。

1921年，"像造汽车一样造房子"的概念就被法国现代主义建筑大师柯布西耶在《走向新建筑》中首次提起，百年后的今天，"像造汽车一样造房子""流水线上生产房子"的装配式建筑建造在关键时间点上得到了完美的演绎。

1.3　结构合理，设计可靠

1.3.1　设计原则

由于建设时机特殊，应急医院的结构设计不仅要保证安全性，更关键的是要在极短的时间内完成建成目标。装配式建设的方式可以有效地压缩建筑的建设周期，采用轻型模块化体系可以摆脱传统建筑建设方式对施工场地的要

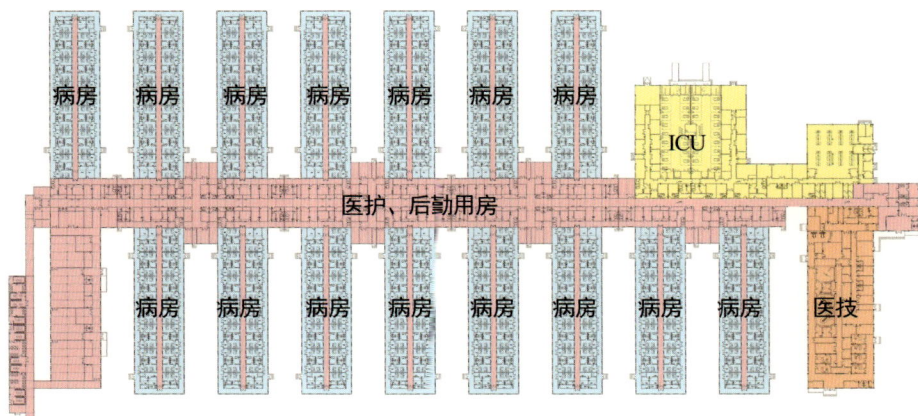

图 1.35　隔离医疗区（北侧区域）平面布置

求，同时满足应急医院功能空间的需求，在应对突发性公共卫生事件的层面具有至关重要的意义 [5]。在 SARS 爆发期间，仅用 8 天时间建设完成的小汤山医院就是典型的采用装配式建设应对突发事件的案例。

雷神山医院的装配式结构设计遵循标准化、模数化与集成化原则，尽可能地利用成熟的工业化产品体系。装配式的模块集成度越高，现场安装工作相对越少，则施工速度越快，成品质量也更易于确保。在结构设计过程中需要充分考虑现场施工条件，设计初期必须与施工方就工期、加工运输、人力设备、材料供应、现场施工方法等方面取得沟通。在充分论证的前提下提出结构方案，确保现场实施的可行性。

1.3.2　上部结构

雷神山医院工程针对隔离医疗区、医护生活区不同建筑功能和空间特点，相应选用了不同的结构形式。

1. 隔离医疗区

隔离医疗区从平面功能和空间特点上可分为病区护理区和医技区两种典型区域，平面布置如图 1.35 所示。其中病区护理单元为规格统一的病房单元与医护办公单元，具有典型的标准化与模块化特征，病区护理单元平面如图 1.36

所示。该单元选择采用轻型模块化钢结构组合房屋（箱式房）结构体系。该体系是指在工厂内制作完成，或者在现场拼装完成，且具有使用功能的轻型钢结构建筑模块单元，可通过装配连接成低、多层建筑。组合房屋的模块单元由模块地板、顶板及墙板组成，同时可将设备管线、门窗及装饰部件集成在模块中。模块单元除满足各项建筑性能要求外，还需满足吊装运输的性能要求 [6]。图 1.37 为病区护理单元箱式房安装现场实景。

组合房屋的模块单元整体采用钢结构骨架和彩钢复合板墙体，骨架以冷弯薄壁型材为主要材料，通过焊接连接，结构整体性强、承载力高。模块各构件组成如图 1.38 所示。模块单元能根据使用需求进行多元化改造，自由拼接，以箱体为基本单元，可单独使用，也可拆除墙板，通过水平及竖直方向的不同组合形成宽敞的使用空

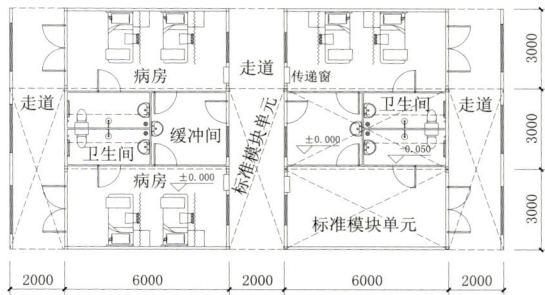

图 1.36　病区护理单元平面图

图 1.37　病区护理单元箱式房现场安装实景

来源：武汉雷神山医院总体建设进度完成 40%[N/OL]. 肖艺九，摄. 新华网，2020-0□-3)[2022-01-28]. http://www.xinhuanet.com/photo/2020-01/30/c_1125514210_6.htm

图 1.39　某箱式房底板结构平面布置

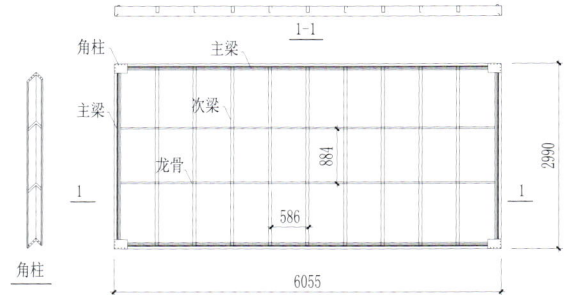

1—上吊角件　　5—顶框保温棉　　9—地板　　　13—下吊角件
2—顶框架　　　6—顶框方管　　　10—地板革　　14—墙体
3—角柱　　　　7—室内吊顶板　　11—窗　　　　15—底框
4—屋顶蒙皮　　8—底框方管　　　12—门

图 1.38　箱式房模块构件组成

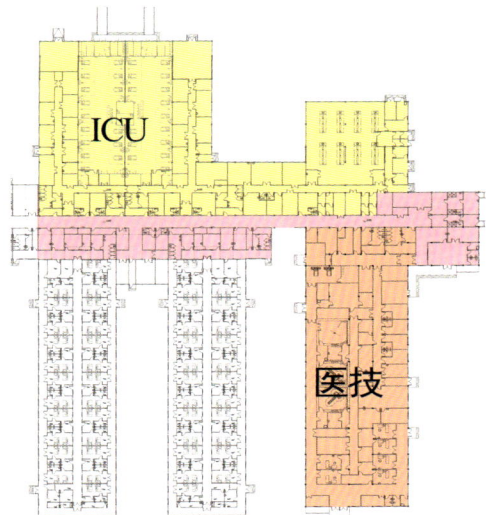

图 1.40　ICU 及医技区局部平面（一期）示意图

间。在垂直方向上可以叠层，一般不超过 3 层。

考虑到箱式房厂家当时的生产及库存情况，选用了名义尺寸为 3.0m×6.0m×2.9m、2.0m×6.0m×2.9m 两种规格的模块单元，建筑平面按照该基本单元模数进行布置。箱式房的整体设计，包括结构构件深化设计均由生产厂家完成，能够满足规范使用荷载的要求。各厂家的产品实际尺寸略有差别，所采用的工艺、材料、截面形式和规格也略有不同。雷神山医院工程采用的其中一种箱式房单元的底板结构平面布置如图 1.39 所示，主要构件参数见表 1.1，结构构件均为镀锌冷弯薄壁型钢，材质均为 Q235B。

隔离医疗区的 ICU 及医技区层高为 4.3m，平面柱网不均匀且跨度较大，一期 ICU 最大柱

距为 7.28m（图 1.40），二期 ICU 最大柱距则达到 18.28m。箱式房、板式房等预制装配式体系难以满足要求，故主体结构采用了钢框架体系。为了便于快速安装，围护墙板与内隔墙仍采用彩钢夹芯板材，柱网尺寸遵循 1820mm 的模数。屋盖采用立缝咬合式面板保温屋面系统。

为了缩短设计周期及深化流程，医技区钢结构设计采用一阶段正向设计方式，即设计一开始即采用详图设计软件 Tekla Structure 进行建模，然后将模型转换至通用有限元软件 Midas/Gen 计算，完成构件截面设计。设计图纸采用模型直接生成加工详图的方式表达，可直接与加工厂对接进行下料并加工。

参照规范对临时结构的要求，医技区钢结

表 1.1

参数类型	部位及描述	
模块尺寸 （长 × 宽 × 高）/mm	外形：6055 × 2990 × 2896 净空：5875 × 2810 × 2575	
结构构件规格 /mm	角柱	L210 × 150 × 3.0 × 3.0，异形折弯件
	屋面主梁	L160 × 100 × 3.0 × 3.0，异形折弯件
	屋面次梁	□ 60 × 40 × 1.4 × 1.4
	底板主梁	L160 × 100 × 3.0 × 3.0，异形折弯件
	地面次梁	□ 120 × 60 × 1.4 × 1.4
外围护体系 规格及材质 /mm	屋面板	0.5 厚镀铝锌彩钢板，带保温层
	地面 装饰面层	2.0 厚 PVC 地板
	地面 基板	18 厚水泥纤维板
	侧墙板	50 厚彩钢玻璃丝棉夹芯板

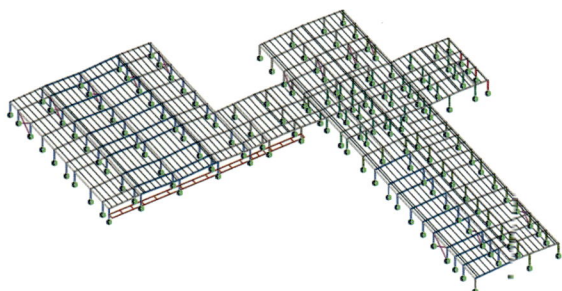

图 1.41 医技区结构形式示意（Midas 分析模型）

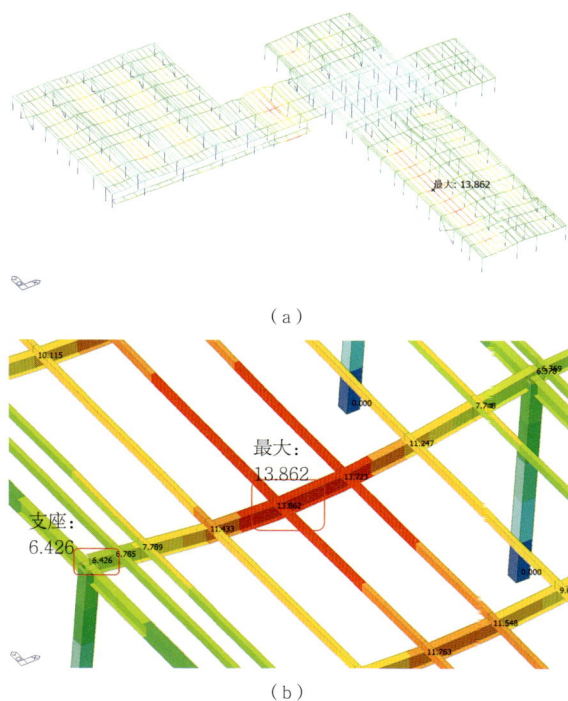

（a）

（b）

图 1.42 标准荷载组合包络工况下竖向位移
（a）整体竖向位移图（mm）；（b）局部竖向位移图（mm）

构设计使用年限为 5 年，结构按承载力极限状态进行设计，并满足正常使用极限状态下的变形要求。屋面恒载取 0.3kN/m²，屋面活荷载取 1.0kN/m²。考虑到轻钢结构体系对风荷载较敏感，风荷载采用了 50 年重现期基本风压，基本风压 w_0=0.35kN/m²，未考虑地震作用。以医技区（一期）的设计为例，给出主要计算分析结果，分析模型如图 1.41 所示。

由图 1.42 可知，在标准荷载组合包络工况下，跨中最大挠度值 d=13.862mm-6.426mm=7.436mm<7280mm/400=18.2mm，满足规范限值要求。

由图 1.43 可知，荷载基本组合下考虑稳定的构件承载力验算结果，梁最大组合应力比为 0.918，柱最大组合应力比为 0.3，均小于 1，满足承载力要求。其中框架柱计算长度系数按有侧移体系考虑。计算屋面主梁稳定性时，考虑檩条

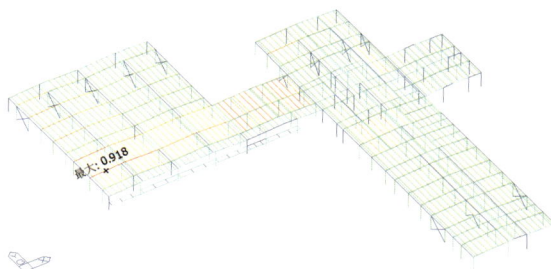

图 1.43 考虑稳定的构件组合应力比

构件类型	构件截面 /mm	型材类型
钢框架柱	□ 300 × 300 × 8 × 8 □ 300 × 300 × 10 × 10	直缝焊接方管
柱间支撑	□ 200 × 200 × 8 × 8	直缝焊接方管
框架梁、非框架梁	HN396 × 199 × 7 × 11 HN350 × 175 × 7 × 11	热轧 H 型钢
主檩条	□ 200 × 100 × 2.75 × 2.75	直缝焊接方管

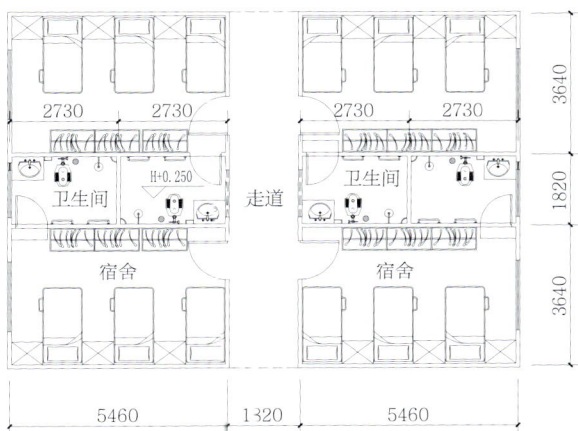

图 1.44　医护生活区典型单元建筑平面

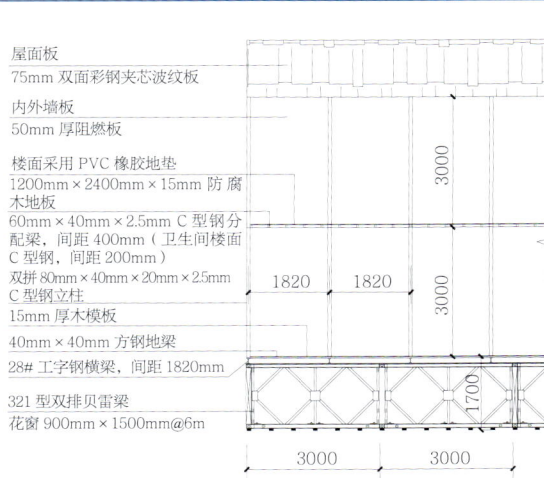

图 1.45　医护生活区活动板房立面构造

侧向支撑的有利作用，主梁平面外计算长度按 4 倍檩距取值，平面内计算长度按梁跨度取值。主要构件材质均为 Q235E，截面尺寸如表 1.2 所示。

2. 医护生活区

医护生活区结构为 2 层，层高均为 3.0m，下设 1.7m 高架空层。根据建筑平面和空间要求，采用了轻钢活动板房体系。建筑平面布置以墙板宽度 1820mm 为模数，标准房间单元平面尺寸为 3640mm × 5460mm，医护生活区典型单元建筑平面如图 1.44 所示。

轻钢活动板房平面布局较箱式房更灵活，能满足医护生活区建筑的各项功能性需求。活动板房技术体系成熟，采用标准化、模数化生产，安装和拆卸较方便，作为临时建筑具有优势。活动板房主体结构为轻型钢框架，承重构件采用冷弯薄壁型钢，构件为工厂预制现场组装，可多次重复使用。医护生活区活动板房立面构造如图 1.45 所示。楼面承重体系采用轻型钢桁架，

其上铺设分配梁与木地板。屋面及外围护墙板采用彩钢夹芯板。活动板房的总体刚度较弱，纵向柱间支撑、横向柱间支撑、楼面和屋面水平支撑是保障其整体稳定的必要措施[7]。

1.3.3　基础设计

1. 岩土工程条件

场地地貌单元为长江三级阶地垄岗与坳沟相间分布区。除部分人工填土外，场地土层主要为第四系冲洪积成因的黏性土及残积黏性土，下伏基岩为白垩—第三系（K-E）泥质粉砂岩，现状场地典型地质剖面如图 1.46 所示。其中粉质黏土层（Q_3^{al+pl}）适合作为天然基础持力层，其承载力特征值 f_{ak}=400kPa，压缩模量 E_{s1-2}=15.0MPa；层顶埋深 0 ~ 11.7m，层厚 4.0 ~ 12.7m；该层在拟建场地大部分地段分布，土质均匀，呈黄褐色、褐红色，硬塑状态，含大量铁锰质氧化物、结核及少量灰白色条带状高岭土，属中偏低压缩性土。

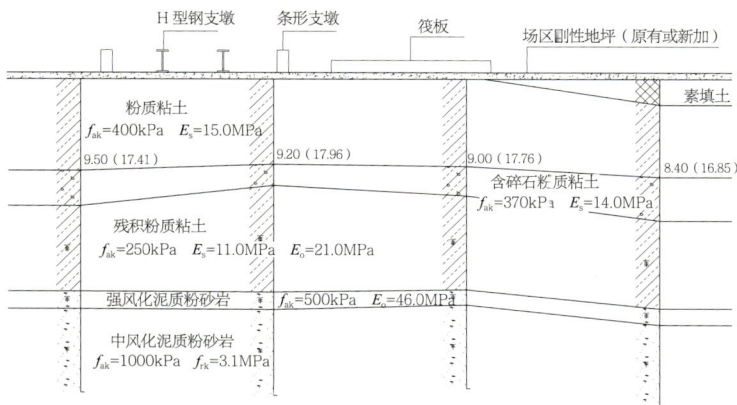

图 1.46　现状场地典型地质剖面图

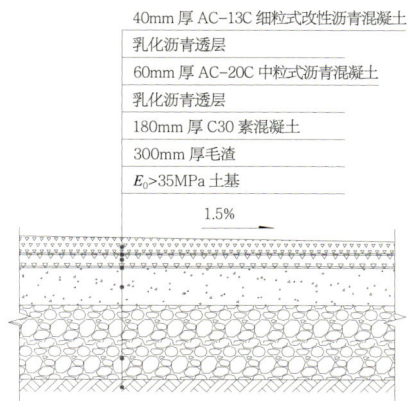

图 1.47　停车场区域地面构造做法

图 1.48　隔离医疗区基础施工现场实景

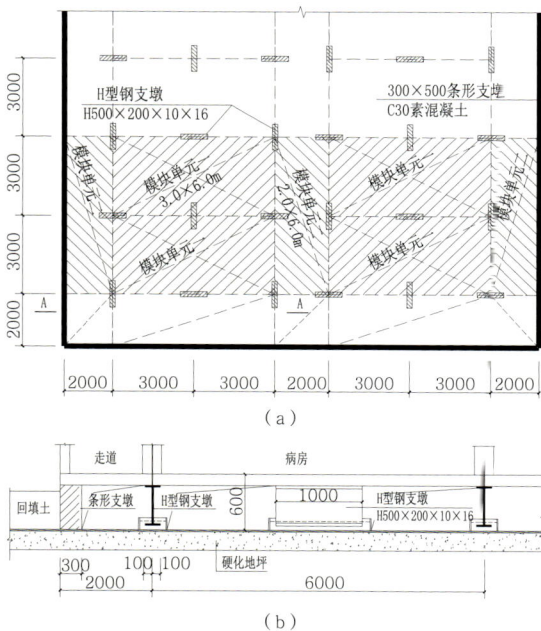

图 1.49　病区护理单元典型区域基础布置图
（a）基础典型平面布置图；（b）A—A剖面

2. 隔离区基础

隔离区选址在原停车场或绿化带区域，其中原停车场区域存在结构硬化地坪，构造如图 1.47 所示。

设计中尽量利用原有硬化地坪。在缺少硬化地坪的绿化区域，复核土层承载力特征值，至少需要达到 60kPa，否则换填处理为砂石垫层。随后现浇 200mm 厚 C30 混凝土硬化层，内置单层双向Φ12@200 钢筋网。先将整个场区的硬化地坪形成整体，后期再根据上部建筑布置来做支墩或筏板。一方面解决了结构基础或筏板的垫层问题，另一方面也为满场区铺设 HDPE 防渗膜（高密度聚乙烯膜）创造了良好条件。施工过程中须提前与给水排水专业协调配合，明确排水主管的走向和标高，确定开挖范围，先开槽、回填，再进行场地硬化施工。隔离医疗区基础施工现场实景如图 1.48 所示。

病区护理单元箱式房自重较轻，硬化地坪可直接承担上部荷载。为满足排水管道安装空间需要，在硬化地坪上设置了一定高度的支墩将箱式房架起。混凝土条形支墩沿建筑外边缘设置，支承上部建筑的同时用于挡住室外回填土，分段式 H 形钢支墩则用于内部范围。病区护理单元典型区域基础布置如图 1.49 所示。

医技区域采用钢框架结构体系，柱底反力较大。若采用硬化地坪难以满足局部受力要求，故在硬化地坪上整浇300mm厚的钢筋混凝土叠合层，形成叠合平板式筏形基础，其具体构造如图1.50所示，其施工现场实景如图1.51所示。

平板式筏形基础自身刚度大，能有效扩散上部结构荷载，降低地基承载力需求，针对局部软弱的不均匀地基有协调沉降的作用。采用平板式筏形基础，使得钢柱脚的位置可以适当调整，摆脱了上部柱网不稳定对基础施工的制约，

极大方便了施工，加快了工程进度。同时还解决了底层架空地板安装高度过大的问题。钢柱脚、架空地板与筏形基础的关系如图1.52所示。

3. 医护生活区基础

医护休息用房位于原万人食堂建筑内部，原有硬化地坪承载力能满足上部2层轻钢活动板房的荷载要求。由于板房下方须留出排水管道安装空间，为避免破除原有地坪，设置了1.7m高的架空层。架空层原设计拟采用钢框架或混凝土支墩，后根据施工单位能够调配的资源，替换为成品321型贝雷梁，形成装配式架空层结构，既满足受力要求，又可循环利用，节约材料，同时也节省了人力，缩短了工期，具体形式如图1.45所示。

1.3.4 结构选型分析

1. 装配式体系的分析

雷神山医院工程根据病区护理单元、医护生活区与医技区的建筑空间需求和功能特点，采用了三种不同的装配式结构体系，各体系的适用条件和体系特点详见表1.3。

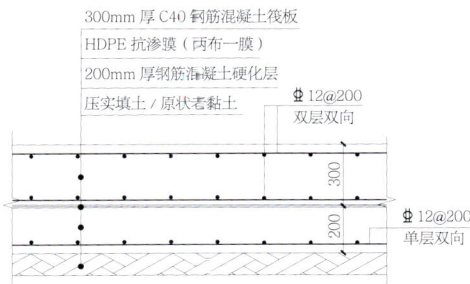

图1.50 叠合平板式筏形基础具体构造

图1.51 平板式筏形基础施工现场实景

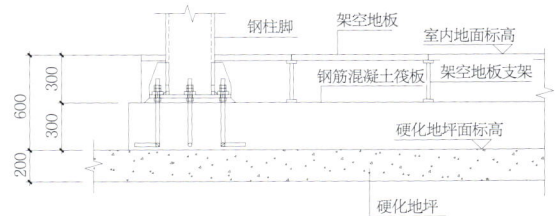

图1.52 钢柱脚、架空地板与筏形基础的关系

装配式体系的适用条件及体系特点   表1.3

结构体系	应用区域	适用条件	体系特点
箱式房	病区护理单元	①建筑具有典型的标准化、模块化的特征；②开间、进深及层高尺寸受限，短期可采购的成品规格较少；③用于短期临时建筑；④可叠放，一般不超过3层；⑤不可承受较大使用活荷载或吊挂较重设备	①采用冷弯薄壁钢构件；②整体均采用工业化生产，集成度高；③不需二次装修，效果较好；④整体吊装，安装速度快，质量易保证；⑤自重轻，整体性好，基础要求低，适应性好
轻钢活动板房	医护生活区	①建筑符合板材模数，具有一定的标准化特征；②开间、进深及层高尺寸受限制，但可选范围比箱式房大；③用于短期临时建筑；④不超过2层；⑤不可承受较大使用活荷载或吊挂较重设备	①采用冷弯薄壁钢构件，构件体系成熟完整；②现场拼装，集成度低；③效果不如箱式房，或需二次装修；④现场拼装速度快，质量易保证；⑤自重轻，整体性好，基础要求低，适应性好
钢框架+轻型墙板	医技区	建筑空间、功能、荷载和使用年限均按需设计，没有限制	①构件可采用各种类型钢材，需设计定制；②各构件及围护体系均现场拼装，集成度低；③需二次装修；④构件加工较复杂，现场焊接较慢且质量不易保证；⑤自重较轻，整体性较好，基础要求较高

名称	建设规模	场地条件及基础形式	结构形式
北京小汤山医院二部工程	总建筑面积2.5万平方米，总床位数612张	采用条形基础	均为单层，采用装配式、标准化单元，整体式钢筋混凝土盒式房屋，部分采用轻钢结构板房
武汉火神山医院	总建筑面积3.39万平方米，总床位数1000张	场区相对高差较大，存在一定面积填方区，局部存在软弱土，采用碎石换填处理，采用筏形基础	1号病区单层，2号病区2层，采用装配式、标准化单元，轻型钢结构模块化箱式房屋与钢框架+轻型墙板体系
武汉雷神山医院	总建筑面积7.9万平方米，总床位数1500张，容纳2000人配套医护用房	场区土质条件较好，具有部分钢筋混凝土地坪可利用，采用条形基础、支墩和筏形基础等多种基础形式	隔离区为单层，医护区生活区为2层；采用装配式、标准化单元，隔离区采用轻型钢结构模块化箱式房屋与钢框架+轻型墙板体系，医护区采用轻钢活动板房

（a）

（b）

（c）

图1.53 部分典型施工阶段实景
（a）场地平整及管槽开挖、预留预埋；（b）基础（支墩）施工；
（c）箱房及钢结构安装过程
来源：湖北日报全媒体记者梅涛 摄

2. 与同类项目的比较

雷神山医院与北京小汤山医院[8]及与雷神山医院同期建设的火神山医院在项目建设规模、场地条件、结构形式等方面的对比详见表1.4。

由表1.4可知，雷神山医院秉承北京小汤山医院的设计思想，同样采用了装配式体系以保证快速建造，但在建设规模、功能、技术与材料等方面更进一步，所采用的装配式体系工业化程度更高。

1.3.5 施工配合

1. 施工过程

雷神山医院的整个施工过程仅用十余天，从2020年1月25日开工至2月6日竣工移交，具体施工过程大致分为如下阶段：①场地平整→②管槽开挖、局部回填→③场地硬化、地坪施工→④HDPE防渗膜施工→⑤条形及型钢支墩、筏形基础施工→⑥箱房安装、钢结构安装→⑦围护墙板隔断及屋面安装→⑧管线安装、装饰装修→⑨医用设备、家具安装及调试。部分典型施工阶段实景如图1.53所示。

2. 施工配合的关键问题

结构设计时须结合施工单位现有资源，采用现有的材料及规格。在充分了解现场施工条件与状态的前提下，提出设计方案，并确保现场能落实。节点构造形式应简单可靠，便于加工，连接方式应便于现场安装。

当施工条件发生变化，按原设计实施有困难时，需要及时配合调整方案。如最初设计方案采用现浇钢筋混凝土支墩，但根据现场人工和材料情况，及时将基础形式调整为 H 型钢支墩。为了解决型钢不便于找平的问题，采取了先用砖或钢板定位，后浇筑混凝土的方法，H 型钢支墩找平构造如图 1.54 所示。

在快速施工过程中，各项工序难免压缩到一起，带来新的问题，如混凝土龄期还未达到，就需要承担上部结构的施工荷载。为弥补混凝土的强度，施工时采用了 C40 早强混凝土，同时在柱脚等受力集中的部位，增配了局部抗剪、抗冲切钢筋。施工现场难免会出现有一些构件漏做、工序不到位的情况，需要随时灵活处置，采取加强措施，以既保证临时结构安全，又不影响工期为原则。

1.3.6 设计体会

雷神山医院作为特殊时期的应急临时医疗设施，设计中面临任务重、时间紧、要求高的挑战，采用了边设计、边施工的特殊工作方式。得益于高效的设计与施工组织协调能力，以及目前较成熟的轻型钢结构装配式、模块化建筑工业体系，在极短的时间内顺利完成了雷神山医院的建设。结合在设计及施工配合中的体会，有以下几点建议供参考：

（1）设计初期需和施工方在工期、加工运输、人力、设备、材料供应、现场施工方法等方面进行充分沟通，减少现场调整和修改方案的情况。

（2）临时医院结构设计优先选择预制装配式结构形式，遵循标准化、模数化、集成化设计原则，要了解和善于利用现有的工业化建筑产品体系。装配式建筑是建造方式的改变，更是设计理念的升级。

（3）要简化设计流程，同步配合施工。

（4）采用适应性强的结构及基础形式，以结构"不变"应方案之"万变"。

（5）结构设计要充分考虑机电管线及设备

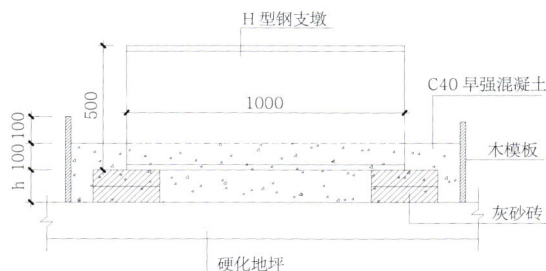

图 1.54 H 型钢支墩找平构造

的空间与荷载要求，并精细配合，以满足给排水的需求。

（6）节点构造形式应简单可靠，便于加工，连接方式应便于现场安装。

（7）应加强模块拼缝处的防水构造，确保金属屋面的抗风措施实施到位。

（8）快速施工过程中应注意安全风险的控制，须采取合适的应对措施。

1.4 因地制宜，灵活选材，优化给水排水方案

雷神山医院的设计在满足应急医院设计要求的基础上，给水排水方案比选及设备材料选择还需考虑市场采购、产品库存、施工便捷、厂商捐赠情况等因素，以最短时间完成设计和施工任务。

1.4.1 项目概况

1. 工程规模

武汉雷神山医院选址于江夏区军运村强军路以北地块，建设用地面积约 22 万平方米，总建筑面积约 7.9 万平方米。该用地西侧地块原规划为军运村食堂，东侧为集中停车场地。

整体规划按照相应标准设计，用于收治已确诊的新冠肺炎患者。根据用地情况将东西两区分别规划为隔离医疗区、医护生活区、保障功能区。项目总床位数约为 1500 张，可容纳医

分区	名称	总建筑面积	建筑高度	层数
医护生活区	宿舍区（共10栋）	14024m²	7.5m（1栋专家楼4.5m）	2层（1栋专家楼1层）
	办公区	1111m²	5.9m	1层
	餐饮区	8889m²	8.0m	1层
	清洁用品库	4462m²	4.3m	1层
隔离医疗区	医疗区	52200m²	ICU：4.3m 其余：2.9m	1层

护人员约2300人。医护生活区、隔离医疗区建筑物见表1.5。

2. 设计内容

给水排水设计涉及室外给水排水消防设计、室内给水及饮用水设计、室内排水设计、室内热水设计、室内消防设计、医院污水处理等。

1.4.2 室外排水方案确定

雷神山医院与火神山医院建设最大的不同是：雷神山医院是建设在原有军运会大型停车场和军运村食堂上，原场地已有300mm厚的混凝土硬化地面，本次建设不需要对场地进行平整，可以直接施工建设。

在室外管网设计过程中，以下问题需要重点解决：

1.4.2.1 根据现场施工进度分步提供图纸

为配合室外排水管网施工进度，室外排水管网的供图时间细分为4个阶段（24h内完成）：①管道沟槽开挖图（便于施工方用破路机等设备破除硬化路面）及室外排水管网走向方案图（让施工方明白设计方的排水管网方案，并现场勘查是否可行）；②室外排水管道管径图（便于施工方备料）；③室外排水管网控制点标高图（便于施工方开挖）；④室外管网完整图纸（第1版）。从设计到竣工这短短的十余天内，共提供了10版给水排水专业室外排水管网图，基本都是根据现场施工条件不断调整的。

1.4.2.2 建筑单体0.000绝对标高的确定

建筑单体0.000绝对标高的确定也是影响室外排水管网施工进度的关键一环。雷神山医院项目分期建设、分期使用（前后使用时间仅相差2~3天），但污水处理站只有1个，室外排水管网需要统筹考虑，一次施工。设计方需要确认现状场地标高是否与原设计图纸相符，分别在项目一期和二期的施工场地内选取9个点，施工方连夜测量并提供数据，设计方第一时间确定建筑0.000绝对标高，确定排水管网最不利点检查井井底的绝对标高。

1.4.2.3 室外排水管网方案确定

雷神山医院项目建设用地面积约22万平方米，根据用地情况，将东西两区分别规划为隔离医疗区和医护生活区（东区少量用地用于建设保障功能区），每区用地面积约11万平方米。

隔离医疗区建设在原有军运村大型停车场上，原场地已有300mm厚的混凝土硬化地面［图1.55（a）］。医护生活区包括在原军运会万人食堂内新建的7栋建筑，以及在室外场地上新建的3栋建筑［图1.55（b）］。

1. 隔离医疗区

（1）管道沟槽的开挖

为保证施工进度，施工方提出要尽量减少场地开挖和回填土方量，同时由于隔离医疗区已有300mm厚的混凝土硬化地面，如全部破坏此硬化路面，施工进度不允许。

（a）

（b）

图 1.55 隔离医疗区及医护生活区实景图
（a）隔离医疗区实景图；（b）医护生活区实景图

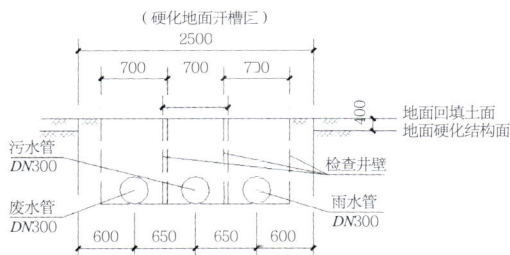

图 1.56 硬化地面开槽区排水管道布置剖面图

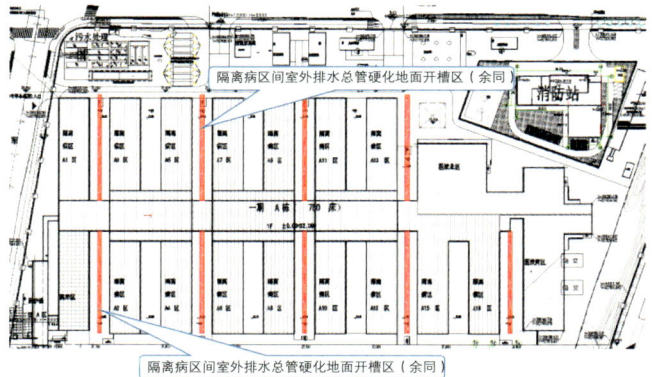

图 1.57 隔离医疗区北区室外排水总管硬化地面开槽区示意图

为满足室外排水管道的敷设需求，并最大限度地减少施工方的开挖量，设计中将所有排水出户管敷设在硬化场地上回填的 400mm 厚的垫层内；隔离医疗区各单元之间的室外雨水、污水、废水总管敷设在 2.5m 宽的开挖沟槽内，且各管道坡度一致，管底标高相同，利于开挖和施工；其余各主干道上的雨水、污水、废水干管也并排敷设在开挖沟槽内（图 1.56），以减少开挖面。在施工过程中，为减少管槽开挖数量，2 个单元共用 1 根室外排水主管，将隔离医疗区各单元之间室外管槽开挖数量减少一半（图 1.57），大大压缩了室外管网施工时间，但会造成室内排水支管过长，设计上考虑在室内设置环形通气管来解决此问题。

（2）排水体制

隔离医疗区室外病区污废水、非病区污废水、室外雨水独立设置排水管网。

（3）管道材料选型

室外排水管材选用PE（聚乙烯）实壁管，采用热熔连接。管道基础采用100mm厚的C15混凝土垫层，上面敷设150mm厚的细砂。

管径不大于600mm，采用成品塑料检查井；管径大于600mm，采用预制钢筋混凝土检查井。井盖采用密封井盖。

（4）污水、废水处理

污水、废水分别排至预消毒接触池进行消毒处理，再经化粪池处理后排至医院污水处理站，处理后达到《医疗机构水污染物排放标准》（GB 18466—2005）中规定的传染病、结核病医疗机构水污染物排放标准，再排至市政污水管网。

（5）室外雨水排放

室外雨水设计重现期为3年，雨水最终排至场地东北侧的埋地雨水调蓄池（有效容积1000m³），调蓄并消毒达标后的雨水排至市政污水管网。

2. 医护生活区

（1）医护生活区室外排水采用雨污分流制、室内采用污废合流制。污水经管道收集，再经化粪池处理后排至市政污水管网，雨水经管道收集后排至市政雨水管网。

（2）在原军运会万人食堂内的医护生活区的宿舍楼，直接建在成品贝雷架上。采用此方

法不仅施工方便、速度快，也解决了排水管道的敷设问题。室外排水尽量利用原军运会万人食堂排水管网和化粪池（化粪池经复核，满足宿舍排水需求），仅局部加设检查井和排水管道。

（3）新建临时建筑未改变原有场地雨水排水系统，故不另设室外雨水管网及雨水口。

1.4.3 给水及饮用水系统

1. 给水及饮用水系统水量保证措施

（1）路市政给水水源保障

医护生活区：在强军路北侧有1处DN200给水接入口，在军体路西侧有1处DN200给水接入口。

隔离医疗区：在强军路北侧有1处DN300给水接入口，在军运路南侧有1处DN200给水接入口。原项目设计从强军路引入1路市政供水管道至生活水箱进口处，为保障水箱进水可靠，从军体路另接入1路给水管道至生活水箱进口处，保证生活水箱进水采用2路市政水源供给（图1.58）。

（2）生活水箱储水容积保障

雷神山医院项目病床总床位数建设目标为1500张，可容纳医护人员约2300人。最高日用水量为1140m³，生活水箱取20%最高日用水量，为228m³，项目原有生活水箱容积300m³，满足建成后的需求，设计中沿用原有水箱不作修改。

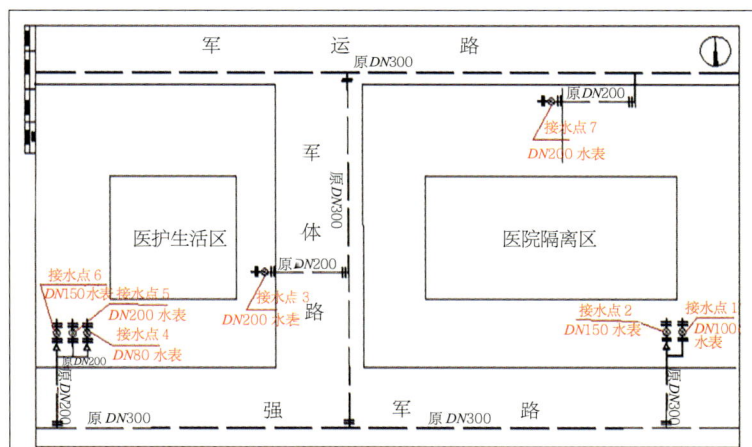

图1.58　市政给水管道及水表点位区位图

（3）加压供水管网环状供水

经计算，原有生活变频供水泵组水量不满足雷神山医院项目要求，设计将原生活供水主泵更换为：$Q=55m^3/h$，$H=35m$，$N=11kW$，4台，4用，并设置一对一变频器。原有辅泵不变。

医护生活区和隔离医疗区生活加压给水均为环状管网供水。室外给水干管设计为$DN200$。

（4）给水管网应急措施

雷神山医院项目对供水可靠性要求较高，考虑水泵房停电等不可抗力因素，加压供水管网设计采用可拆卸金属波纹管方式与市政供水管网相接。此部分金属波纹管采用法兰连接，平时拆除，且波纹管前后阀门关闭。在应急时安装金属波纹管并打开前后阀门。《城镇给水排水技术规范》（GB 50788—2012）条文说明第3.4.7条指出："《城市供水条例》中明确：'禁止擅自将自建设施供水管网系统与城市公共供水管网系统连接；因特殊情况需要连接的，必须经城市自来水供水企业同意，报城市供水行政管理部门和卫生行政主管部门批准，并在管道连接处采取必要的防护措施。'"

施工中，因产品采购紧急，加压供水管网采用止回阀与市政供水管网相连，项目给水方案报水务部门批准。

2. 给水及饮用水系统水质保证措施

雷神山医院项目供水全部采用断流水箱供水。生活水箱一体化设备位于医护生活区室外，泵房内出水管处设RZ-UV2-DH200FW型紫外线协同防污消毒器，如图1.59所示。

图1.59　一体化生活泵房及紫外线协同防污消毒器实景图

隔离医疗区内生活给水按病区和医疗区分别设置给水管网，在病区给水引入总管前端设置倒流防止器。在建筑物内，各病区给水按护理单元分设控制阀门，阀门设置于清洁区易于操作处。

考虑雷神山医院项目为应急医院，生活供水系统设置应急加氯装置。在生活加压泵房生活水箱进水管处预留$DN32$投加口，并预留计量泵，用于应急加氯。

3. 给水及饮用水系统水压保证措施

雷神山医院项目生活给水采用变频加压供水泵组供水。由于给水管路较长，室外采用$DN200$环状管网，室内采用$DN150$给水干管，所有给水干管不变径。

4. 洁具选型

医护人员使用的洗手盆，细菌检验科设置的洗涤池、化验盆，以及公共卫生间洗手盆采用感应式水龙头。应国家卫生健康委要求，医护生活区、隔离医疗区医生卫生间采用蹲便器，隔离医疗区患者卫生间采用坐便器；蹲便器采用脚踏式冲洗阀。

5. 饮用水选择

医护生活区及隔离医疗区均分散设置开水间，每个开水间内配置2台12kW开水炉。开水炉带过滤功能，可提供开水、常温水。

1.4.4　室内排水系统

1.4.4.1　隔离医疗区

1. 排水体制

隔离医疗区病区污废水、非病区污废水分流排放，采用各自独立的排水管道。

2. 排水管道水封保障措施

（1）排水出户管距离较长，在各排水出户管末端设环形通气管，以保障管道排水能力，并防止水封被破坏。

（2）准备间、污洗间、卫生间、浴室、空调机房等设置地漏，护士室、治疗室、诊室、检验科、医生办公室等房间不设地漏。地漏采用带过滤网的无水封地漏加存水弯，存水弯的水封高度为50mm；手术室、急诊抢救室等房间的地漏采用可开启的密封地漏。

（3）洗手盆的排水给地漏水封补水。

3. 其他保障措施

（1）每个隔离医疗单元的数个环形通气管汇合成汇合通气管，穿外墙并升至屋面以上，在屋面设置紫外线空气杀菌消毒器以消毒，并且排水通气管的位置避开屋面新风机房的位置。

（2）空调冷凝水间接排至地漏，进入医院污水排水系统。

（3）为保障患者如厕方便及安全，根据国家卫生健康委要求，病房卫生间大便器采用坐便器。

1.4.4.2 医护生活区

1. 排水体制

医护生活区新建建筑主要为医护人员、工作人员宿舍，排水采用污废合流制，设计较为常规。原有建筑使用功能调整后，在充分利用原有室外排水管道的基础上对排水管道进行改造以满足使用需求。

2. 室内排水管网

原食堂大空间内部新建的7栋宿舍楼各卫生间的立管排水由排水横干管汇合。排水横干管安装于由成品贝雷架架高的1.5m高架空层内，并在大空间内人行通道处设排水管道过桥以不影响交通。此方案是在紧急建设情况下设计的合理的排水横干管安装方案，不仅施工方便，且避免开挖原食堂地面。

原食堂大空间内部临时建筑的各排水立管通过汇合通气管后，穿侧墙再伸顶至原食堂屋面以上通气。

1.4.5 室内热水系统

1. 热水设置部位

雷神山医院项目的隔离医疗区和医护生活区都有热水需求，且需求量较大。根据雷神山医院项目的特点及厂商捐赠情况，采用不同的方式提供生活热水。医院生活热水系统如表1.6所示。

2. 热水方案比选

针对雷神山医院的三种热水供应方式，分析其优缺点，如表1.7所示。

3. 热水方案实施分析

设计过程中，很多知名厂商主动与甲方联系，要求免费赠送产品和设备，以表达对武汉人民的支持。能提供热水的设备有电热水器、空气源热泵机组和商用燃气热水炉，它们在使用特性上各有优缺点。雷神山医院项目热水需求量较大，设计需要结合项目的特点，在满足使用功能的同时，充分发挥各种设备的优势（图1.60）。

医院生活热水系统   表1.6

区域	部位	使用部位	电热水器容积	电热水其功率	台数	品牌
隔离医疗区	北区	病房	60L	2.2 kW/台	360台	海尔、美的
		医护	—	采用集中热水，商用燃气热水炉，每台99 kW	2套（1套4台，1套5台，共9台）	史密斯
		集中浴室	—	采用集中热水，商用燃气热水炉，每台99 kW	1套（共4台）	史密斯
	南区	病房	60L	2.2 kW/台	360	海尔、美的
		医护	—	采用集中热水，商用燃气热水炉，每台99 kW	1套（共7台）	史密斯
		集中浴室	—	采用集中热水，商用燃气热水炉，每台99 kW	1套（共4台）	史密斯
医护生活区	医生宿舍1~4号楼	采用集中热水，空气源热泵机组，每日供水量40m³/套（55℃）			1套	美的
	医生宿舍5~7号楼	采用集中热水，空气源热泵机组，每日供水量30m³/套（55℃）			1套	美的
	部队宿舍和专家楼	采用集中热水，空气源热泵机组，每日供水量35m³/套（55℃）			1套	美的

热水供应方式	优点	缺点
电热水器供应热水	病房采用电热水器比较灵活，不会造成交叉感染	①数量太多，需要配电的总功率太大，给变压器造成的压力太大；②由于电热水器容积有限，无法连续使用；③电热水器安装麻烦，集装箱拼装建筑的墙上可以承受容积式电热水器的重量，但是板房的卫生间墙壁承受60L电热水器的重量有困难
空气源热泵机组供应集中热水	①可以做到随到随洗，热水使用有保障；②节能	①管路太长，热损失大；②由于设有热水箱和空气源热泵，室外需要较大场地
商用燃气热水炉供应集中热水	①可以做到随到随洗，热水使用有保障；②燃气炉可以设置在室外，且占地面积小	①管路太长，热损失大；②室外需要有燃气供应；③距离室外变压器需要有4m的安全距离

图 1.60　各区的热水供应实景图
（a）空气源热泵机组（医护生活区）；（b）商用燃气热水炉（隔离医疗区）；（c）电热水器（病房区）

（1）隔离医疗区病房热水供应

由于每间病房最多住2人，病房采用电热水器供应热水有非常明显的优势，可以避免交叉感染。

同时病房区域的建筑为集装箱拼装建筑，病房电热水器容积为60L，功率2.2kW/台，注满水后总重81kg，经复核，卫生间墙壁可以承受81kg的重量。

（2）医护生活区热水供应

医护生活区共有房间398间，除了14间专家单人间，其余均为6人间。根据项目特点，医生工作虽然采用倒班制，但是上下班时间较为集中，且每个房间人数较多，卫生间淋浴同时使用的概率较高。医护生活区建筑为集中板房，卫生间墙体较软，承受81kg的电热水器重量有困难。同时医护生活区室外有宽阔的位置可以放置空气源热泵机组及水箱。虽然医护生活区由之前的军运会万人食堂改造，附近就有燃气，但出于节能考虑，设计采用3套空气源热泵机组为医护生活区提供集中热水。

（3）隔离医疗区热水供应

隔离医疗区在每个医护单元的一更和二更区域均设有多个淋浴间，共有180个淋浴头，同时在南北区分别设有1个集中浴室，共有淋浴间64间。由于隔离医疗区用地紧张，室外没有场地放置空气源热泵机组和水箱，同时隔离医疗区有燃气供应，故设计结合医生使用热水的需求及特性，考虑采用5套商用燃气热水炉集中供应淋浴热水。

1.4.6　消防系统

由于雷神山医院的设计与建设迫在眉睫，且医院为人员密集场所，如何把握医院消防系统设计标准，既满足医院建设的工期要求，又满足医院使用功能需求，同时兼顾医院的消防安全，最大限度降低医院消防安全隐患，阻止火灾进一步蔓延，对给水排水专业消防系统设计提出了挑战。

雷神山医院为临时建筑，医院北侧为江夏区消防救援大队所属的雷神山消防救援站，站内有消防车和消防人员就地待命，可以在第一时间为雷神山医院实施消防救援。项目消防设计方案经与消防主管部门沟通，得到了消防部门的认可。

1. 消防水源的安全保障

保障室外消防水源的安全和完善室外消火栓管网是雷神山医院项目消防系统设计的首要目标，以保证在发生火灾时，消防队员可通过室外消火栓取水达到灭火的目的。

项目周边强军路、军体路、军运路、黄家湖路、黄家湖大道均有完善的市政环状给水管网，市政给水管管径从 $DN300$ 到 $DN600$ 不等，且市政给水压力为 0.35MPa。

医护生活区从强军路和军体路分别接入 1 路 $DN200$ 的市政给水接口；隔离医疗区从强军路和军运路分别接入 1 路 $DN200$ 的市政给水接口。

2. 室外消防管网及室外消火栓的布置

医护生活区和隔离医疗区室外消防管网均为独立管网（与生活管网分开），且呈环状布置

（图 1.61）。室外消防管网上布置有地上式室外消火栓，地上式室外消火栓有 $DN150$ 的栓口 1 个以及 $DN65$ 的栓口 2 个。室外消火栓间距控制在 120m 以内，距离路边不大于 2.0m。

3. 室内消防设施

改建的建筑，充分利用原军运会建筑的室内消防设施，并加强建筑灭火器的布置；新建建筑，由于时间的原因且雷神山医院为临时建筑，医院北侧的消防站有人员及消防设备随时待命，故不设置室内消火栓系统和自动喷水灭火系统，只加强建筑灭火器的布置；对于医院强电间，设置悬挂式超细干粉自动灭火装置，加强保护。

1.4.7 污水处理与消毒

雷神山医院医疗污废水主要为来源于隔离医疗区的生活污水、医疗废水。水中除含有常规的有机物外，还含有新冠肺炎病毒、细菌、化学药剂等，成分较为复杂，直接排放会对周围环境造成潜在的威胁，必须进行有效的处理后方可排放。

污水处理站选址于武汉市军运村 3 号停车场北侧地块，主要处理隔离医疗区的所有污废水，最大日排水量为 1200m³，考虑一定的安全系数，总处理规模按 2 条 40m³/h 处理单元并联运行设计，总规模为 80m³/h。

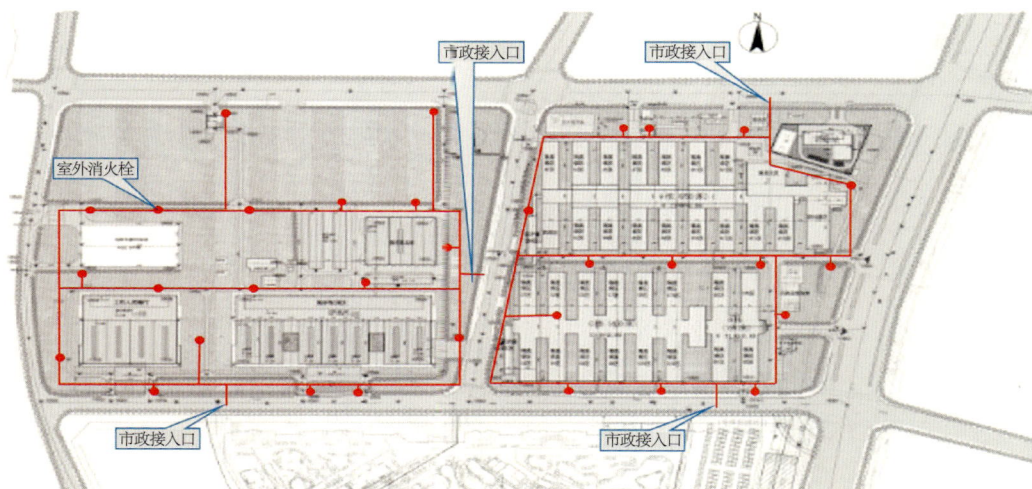

图 1.61　室外消防管网布置图（武汉雷神山医院）

1. 排水收集管网

隔离医疗区与活排水包含病区污废水、非病区污废水，分别设置排水管网单独收集，分别进入预消毒接触池，有效地防止病区排水进入非病区排水管网，阻断气溶胶传播及粪口传播途径，保证医护人员安全。

2. 污水处理工艺

雷神山医院属应急医院，污水处理采用预消毒接触池＋化粪池＋提升泵站（含粉碎格栅）＋调节池＋MBBR生化池＋混凝沉淀池＋折流消毒池工艺，达到《医疗机构水污染物排放标准》（GB 18466—2005）表1传染病、结核病医疗机构水污染物排放限值中的要求后排放。根据《医院污水处理工程技术规范》（HJ 2029—2013）和相关工程经验确定污水处理工艺，如图1.62所示。

3. 消毒工艺

雷神山医院建设时武汉市处于疫情期，项目污水处理的关键之处在于污水二次消毒、预消毒接触池及化粪池通气管废气处理、污水处理站臭气消毒处理。

污水处理工艺中共有两次强化消毒：一次消毒位于预消毒接触池进口，二次消毒位于折流消毒池进口。预消毒接触池须采用密闭储罐，污水停留时间3h，加氯量40mg/L（有效氯）。折流消毒池中污水停留时间1.5h，加氯量25mg/L（有效氯）。

预消毒接触池采用推流式，推流式的水力性能使氯与微生物能最大限度地接触，在有效的消毒时间内达到最充分的混合，取得最有效的消毒效果。预消毒接触池中污水悬浮物及杂质较多，而折板式构造中缝的间隙小，很容易形成堵塞，所以不能采用折板式构造。折流消毒池中的污水经混凝沉淀池处理后，水质较好，可采用折板式构造。

在消毒剂的选用上，对液氯消毒、二氧化氯消毒及次氯酸钠消毒进行了比选。液氯具有持续的消毒作用，适用于远离居住区的规模较大的（大于1000张病床）且管理水平较高的医院污水处理系统。二氧化氯具有强烈的氧化作用，不产生有机氯化物，投放简单方便，不受pH值影响。次氯酸钠适用于规模小于300张病床的、经济欠发达地区的医院污水处理系统。经过多次专家论证，为加强消毒效果，采用液氯消毒的消毒方式。

在建设的过程中，一方面，依据公安部颁布的《剧毒化学品、放射源存放场所治安防范要求》（GA 1002—2012），液氯的存放与使用要求极为严格，使用不当会造成爆炸，液氯加药间构筑物的防爆抗爆、防火间距等均有要求，施工工期长；另一方面，现场已备有液氯投加机、泄氯检测及报警装置和泄氯吸收装置，但液氯氯瓶的采购及运输需要办理很多手续，不能及时到货，项目的交工时间紧，难以满足需

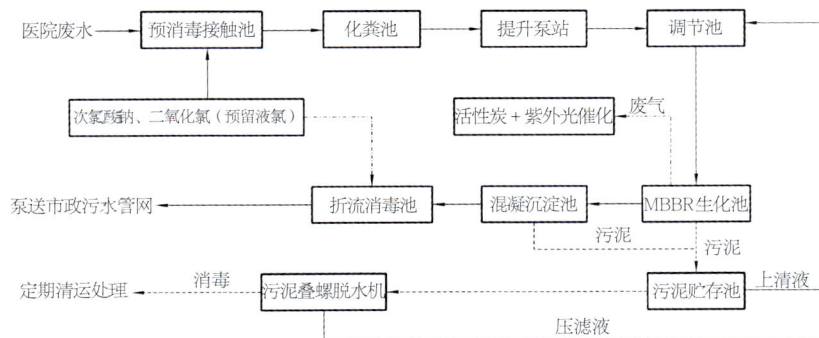

图1.62 污水处理工艺流程图

求。后经与武汉市生态环境局沟通，采用二氧化氯消毒。采用一体化二氧化氯设备，用氯酸钠与盐酸经二氧化氯发生器制备二氧化氯消毒剂。预消毒接触池加氯量 40mg/L，折流消毒池加氯量 25mg/L，设计最大加氯能力为 6.5kg/h。

预消毒接触池及化粪池通气管集中收集废气后于加药间外墙处伸顶，高于加药间屋面 0.5m，通气口顶部设紫外线＋活性炭吸附装置。污水处理站臭气经收集后由紫外线＋活性炭催化消毒处理后排放。为确保过滤吸附效果，须定期更换活性炭高效过滤器。

1.4.8 设计反思

雷神山医院投入使用后运行情况一切良好。从设计到投入使用共十余天时间，设计院、施工方、监理方、代建方、业主、政府职能部门通力协作，畅通无阻。给水排水团队共有 12 人参加，团队成员均为室主任工程师以上级别专业人员，每人分工明确，给水排水总工程师负责确定方案、对内配合、对外沟通、汇报以及现场服务，设计人员负责实施，并由专人负责管理电子版图纸，并下发电子版图纸至施工方，真正做到边设计、边施工。

室内外给水排水方案的确定在满足应急医院建设标准的基础上，同时满足医生和患者使用需求，尽量满足施工方便、快捷的要求，所有材料的选择均考虑市场采购、产品库存、厂商捐赠情况。

1. 重难点分析

（1）室外排水管网的施工在所有项目建设的前期进行，直接影响建设工期，故室外排水方案的选择既要满足应急医院室外排水需求，又要满足医院建设的工期要求，尽量减少场地开挖和回填土方量，压缩室外管网施工时间。

（2）污水处理站消毒方式有很多种，设计针对消毒剂的选用对液氯消毒、二氧化氯消毒

及次氯酸钠消毒进行了比选。本次设计综合比较，最后确定采用二氧化氯消毒。

（3）雷神山医院为临时应急医院，是人员密集场所，如何把握医院消防系统设计标准，对给水排水专业消防系统设计提出了巨大挑战。消防系统设计在消防水源和室外消防管网有保障的前提下，为满足工期要求，室内消防设计不拘泥于国家消防规范，充分利用了原军运会建筑已有的消防设施条件，对已有消防设施进行改造，加强灭火器的布置，同时加强消防管理措施，以保障消防安全。

2. 后期工程回访

疫情基本缓解后，中南院组织雷神山医院设计回访，实践证明，雷神山医院的给水排水设计安全有效，充分满足应急医院的使用需求，但有以下问题需在以后的设计中进一步完善。

（1）污水池和洗手池在选购时应加大一号，防止污水外溅。

（2）阀门的设置位置应便于检修和维护。

（3）ICU 应设置热水装置，满足患者的使用需求。

1.5 电气及智能化

1.5.1 项目背景

2020 年初，为遏制新冠疫情，武汉市迫切需要在最短时间内建立专业化的呼吸类临时传染病医院，中南院因此承接了雷神山医院的建筑设计任务。该医院位于江夏区军运村，建设用地面积约 22 万平方米，总建筑面积约 8 万平方米，包括负压隔离医疗区、医技净化区、医护生活区、污水处理站、氧气站及配套用房等，病床总床位数约 1500 张，可容纳医护人员约 2300 人。

设计伊始，项目组电气专业成立核心团队，查阅了所有的相关设计规范、标准和有关资料，

结合项目的特点，进行设计方案的比选和深入研究，很快就确定了电气的设计范围、设计内容和设计标准，并制定了设计原则、任务分配、日程计划等，同时确定了供配电系统及智能化系统方案。在夜以继日的设计过程中，校对、审核与设计同步进行，保证了设计进度，经过设计团队不分昼夜地辛苦努力，仅用5天时间就圆满完成了设计任务。电气设计历程如下：

2020年1月24日，农历除夕，接到设计任务，所有专业负责人勘查现场，当晚建筑专业开始进行方案设计，电气专业组织设计团队。

2020年1月25日，当晚建筑专业提供医护生活区、隔离医疗区建筑方案图及总平面图，电气专业研究建筑图纸，制定设计原则，确定电气设计方案，包括供配电系统、照明系统、防雷接地系统、智能化系统等所有相关系统。

2020年1月26日，完成电气及智能化系统设计说明，作为设计指导文件，完成医护生活区电气专业全套施工图。

2020年1月28日，与供电部门沟通，明确总平面图中箱式变电站、柴油发电机组的位置。完成室外电气总图、主要设备订货清单。

2020年1月30日，完成隔离病房区及医技净化区电气施工图纸。提供雷神山医院全套电气施工图纸。

2020年2月7日，雷神山医院顺利完成施工验收。

2020年2月8日，雷神山医院开始交付使用，医护人员及患者入住。

在即将完成雷神山医院设计时，2020年1月29日收到了湖北省住房和城乡建设厅关于《呼吸类临时传染病医院设计导则（试行）》（以下简称《导则》）的编制任务。电气设计、智能化设计这两章内容由中南院负责编写，1月31日完成了送审稿，经过专家评审，2月3日对外颁布，以指导同类医院的电气设计。

为总结雷神山医院的电气设计经验及解释《导则》中电气、智能化设计章节的编制思想，后面几节将对该类医院电气及智能化设计的内容、特点、重点和难点进行全面的分析和阐述。

1.5.2 电气设计特点

呼吸类临时传染病医院属于医疗建筑，必须遵守一般医疗建筑的相关规范，同时它也属于传染病医院，还需要遵守传染病医院的相关规范，但它与普通医院有显著的区别，与一般传染病医院相比较也有明显的不同。

首先，它是用于收治呼吸类传染病患者的医院，需要注意以下四点。

（1）它的负压通风设备主要是为了让污染区产生负压，避免对清洁区的空气造成污染，这对于防止呼吸类传染病的病毒在建筑内传播特别重要，必须纳入特别重要负荷的供电范围，除由2路市电供电外，还须归入柴油发电机组的供电范围。

（2）医用气体设备对于肺炎患者而言属于性命攸关设备，制氧及供氧设备、压缩空气及负压吸引设备等属于抢救危重症患者的重要设备，也必须和负压通风设备一样纳入特别重要负荷的供电范畴。

（3）呼吸类传染病的传染性极强，消毒杀菌设备并不局限于布置在洗消及洁净场所，几乎所有的医疗场所都需要考虑设置消毒杀菌设备的电源。空气杀菌主要采用紫外杀菌灯及空气消毒器等。设计中不能忽视各类场所的杀菌器电源设置，除房间内应设置外，走道也须每隔3～4m设置1个紫外杀菌灯电源插座。

（4）呼吸类传染病医院按照"三区两通道"设计，电气管线管口及穿墙孔洞必须实施密封措施，以防止病毒的跨区传播。同一区域的电气管线管口、穿墙孔洞及电气设备的安装缝隙也需要密封，防止病毒进入，难以消毒。同时电气设备应选用表面光洁、易于清洁的产品。

其次，它是临时性医院，为挽救更多患者的生命，需要在最短的时间内建成，因此其设计周期和建设周期都非常短，必须遵循快速、安全、通用的原则。需要在不违背规范强制性条文的大框架下予以创新和改进，有必要做到以下几点。

（1）精选电气及智能化系统设计内容，非必要的和用不着的系统一律不设置，比如能源管理系统、智能照明系统、视频示教系统、信息发布系统、建筑设备监控系统等。但必需的系统不可缺少，比如医护对讲系统等。不设置非必要系统的目的是加快设计进程，减少施工及调试时间。

（2）电气系统必须简洁、可靠，推荐采用模块化设计。比如2台变压器和1台柴油发电机组成1个供电模组，保证高可靠性的同时，也满足设计快速、施工方便的要求。另外，病房区按相似区块成组设计，便于设计套用，也便于组织施工。

（3）电气设备必须满足安全可靠、技术成熟、施工快速、调试方便、货源充足的要求。比如雷神山医院项目采用箱式变电站及箱式静音柴油发电机组，相比室内机房而言，可由工厂提前加工预制，更便于施工和快速就位，检修也方便安全。另外，将变压器容量限制在800kVA及以下，就可以采用简单的熔断器保护，避免复杂的继电保护调试过程，施工快速且供电可靠，同时货源很充足。必须避免选用技术复杂、调试漫长、货源稀少的电气设备产品。另外，疫情期间工厂大多停工，没有库存备货的产品就满足不了施工进度要求。

呼吸类临时传染病医院类似战时医院，需要打破常规，提前制定快速安全的设计原则。如果还是按照传统医院的模式按部就班设计，将会贻误时机，拖延工期，同时可靠性并未提高。如何合适地选择电气方案，如何合理地理解现有规范，是设计师，特别是电气专业负责人必须把握的重要事项。

1.5.3 负荷分级和负荷计算

关于用电负荷分级，可以参照《医疗建筑电气设计规范》（JGJ 312—2013）和《传染病医院建筑设计规范》（GB 50849—2014）这两本规范，不再赘述。但对雷神山医院这样的项目而言，规范中的负荷分级过于细分，如果严格按照规范中的负荷分级来进行分类供电，显然不利于快速设计，系统也会过于复杂。同时，相对于普通的医疗建筑，雷神山医院这样的建筑的重要作用是挽救患者生命，其供电负荷等级明显更高，还有前文谈到的特点，比如，负压通风设备及医用气体设备应纳入特别重要负荷的供电范围，这也超出了规范的内容，所以，在设计中并没有完全按照规范中的负荷分级来设计，而是按照建筑的分区来统一供电负荷等级，从而确定供电方案。

先将容量较大的非重要负荷，如舒适性的电辅加热采暖设备（武汉不属于严寒地区）、病房淋浴电热水器等作为三级非重要负荷，从用电负荷中分离出来。其余用电则按照建筑分区及设备类型来分级供电。具体做法是：将净化手术区、医技区、急诊区、监护病房区、负压隔离病房区（电辅加热采暖设备除外）、检验中心、放射设备、负压通风设备、医用气体设备、污水处理设备、电梯（如果有）、消防用电设备、智能化系统设备等确定为一级负荷中的特别重要负荷，若有中心供应、医用焚烧炉、太平间冰柜等设备时，也宜归入特别重要负荷；其余的用电设备均按照一级负荷的要求供电。这样，既便于快速设计，简化供配电系统，又提高了供电可靠性。

如果单纯按照国家现有规范的规定来设计，可能并不一定合适，例如，医护人员的宿舍按照规范只能作为三级负荷供电，但是，雷神山医院项目的医护人员在一天繁重的工作后需要得到充分的休息，以利于第二天的工作，如果电源不可靠，影响医护人员的休息，将会带来严重的不良后果。建筑中这样的负荷还有很多，提高部分负荷的供电等级，对雷神山医院项目而言，是非常必要的，甚至是必须的，所以，需要按照项目的特点来设计。我们在编制《导则》时，也是基于这一原则。

关于用电负荷计算，由于雷神山医院项目大量采用电采暖、电热水及厨房电炊设备，所以，单位建筑面积的用电负荷比普通医疗建筑要大很多，这是水暖专业为满足项目建设速度，不得不作出的选择。电气设计人员在规划供电方案时不能按照以往的经验来预估供电容量，而是需要按照项目特点作好充分的预估，以便与供电部门接洽，申请足够的供电容量。由于非消防重要负荷容量较大，而发生火灾时需要切断非消防负荷，所以，平时不用的消防负荷不应计入计算容量。

同时，还需要考虑合理的需要系数，不能一味守旧，节能与节省投资也是电气设计需要注意的内容。雷神山医院项目病房及生活区的电热水器不可能都处于同时加热状态；空调供暖设备也会在达到预定温度后就处于保温状态；病房区的医护电淋浴设备也呈现分时段使用的状态；电炊设备的用电高峰与就餐时间密切相关；病房和宿舍用电大部分时间都处于轻载状态；放射设备虽然功率大，但却是瞬时尖峰负荷。采用合理的需要系数，可以节约投资，减少供电部门的增容压力。只有这样，才能得到合理的变压器装机容量，确定供电系统方案。雷神山医院用电负荷的综合需要系数（计入同时系数）经分析计算为 0.6，配置类似设备的医院的需要系数在 0.6 ~ 0.7 比较合理。

在设计前期，预估装机容量对于规划供电方案是非常重要的一环。下面将雷神山医院的设计指标列出来供同类型的建筑电气设计参考。雷神山医院采用了全电辅助供暖、分体空调，病房及生活区采用电热水，厨房采用大量电炊设备，变压器装机容量达到了 17720kVA，负荷密度为 224VA/m²。如果办公更衣区全部采用电热水的话，负荷密度可达到 250VA/m²。

所以，对于同类项目而言，需要知道是否有天然气供应，是否有可利用的热源，由此了解清楚是否需要电辅供暖、电制冷、电热水、电炊设备，从而确定项目的供电负荷密度，便于前期快速确定供电系统方案。不同的项目，其水暖设备的选型不同，变压器装机容量也可能相差较大，需要进行仔细的了解、分析和计算。根据不同的用电设备状况、项目规模，以及项目所在地区的天气和能源供应情况，我们认为同类项目的负荷密度区间定为 150 ~ 300VA/m² 比较合适，此数据可供设计同行参考。雷神山医院共设计有 28 台室外箱式变电站（合计 17720kVA）及 11 台室外箱式柴油发电站（合计 6870kVA）。

1.5.4　供配电系统

可以根据以上两小节的内容来进行供配电系统的规划设计，根据一级负荷的供电要求，应采用 2 路相互独立的 10kV（或 20kV）高压市电电源供电，即来自不同的 110kV 区域变电站或同一变电站由不同的上级电源供电的母线段。由于雷神山医院项目可以理解为急救医院或战时医院，其用电负荷可全部按重要负荷考虑，采用 2 路电源相互备供、100% 备供的方式。雷神山医院供电网络示意图如图 1.63 所示。

雷神山医院由于变压器装机容量较大，而且建于军运会相关设施的原址上，供电条件较好，供电部门采用了 4 路 10kV 供电电源，分成 2 组，每组均为 2 路相互独立的电源，互为 100% 备供。

对于特别重要负荷（见 1.5.3 节内容），除

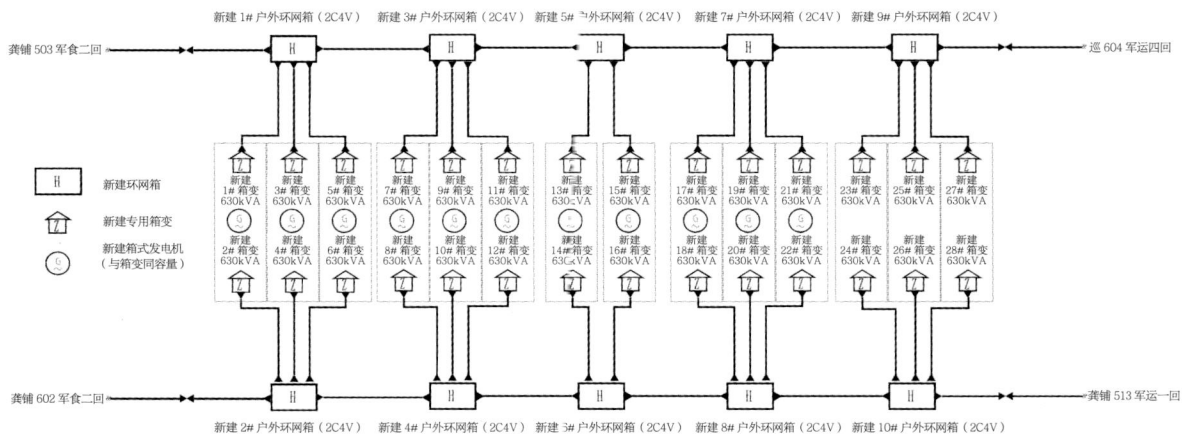

图 1.63 雷神山医院供电网络示意图

采用 2 路 10kV 电源供电外，另设置应急柴油发电机组作为备供电源，发电机组应在市电停电时，15s 内自动启动并输出电能。

为便于快速施工，宜采用室外箱式变电站及室外箱式静音型柴油发电机组，并分区集中设置。发电机组应自带日用油箱，并留有供油接口。当地应有可靠的柴油供应，当市电停电时，可以采用供油车的方式快速供油。若当地无此条件，应考虑设置埋地式储油罐，储油罐的储量须根据当地供电部门保证的停电恢复时间确定。

高压系统采用环网式供电，可以采用两种方式，当箱式变电站的高压柜采用的是 2 进 1 出型环网柜，即进线采用单环网，出线带熔断器保护时，则可以利用箱式变电站组成高压环网；当箱式变电站没有环网柜时，则可以在室外设置高压环网箱，由环网箱向箱式变电站供电。环网箱采用单环网，至变压器出线采用熔断器保护。雷神山医院共设置了 10 座室外环网箱（2C4V型），环网箱为 2 进 4 出型，每座环网箱为 2 台或 3 台箱式变电站供电。

单台变压器及发电机容量不宜大于 630kVA，最大不宜超过 800kVA。可以采用简单可靠的熔断器保护，不必采用断路器和综合继电保护器。变压器应按 2 台 1 组设计，同时工作，

互为备供。成组的 2 台变压器应由相互独立的 2 路高压电源供电。变压器负载率不宜大于 60%。柴油发电机容量宜与变压器相同，按 1 台发电机与 1 组变压器对应配置。其出线宜在变压器低压总开关处自动切换。当受条件限制时也可采用其他合适的方式。供电设备应按区分组，不能交叉供电。

为什么变压器负载率不宜大于 60%？这是因为，当 1 台变压器发生故障时，另外 1 台变压器可利用其长期过载能力（约 120%），承载 2 台变压器的所有负荷输出。也可以按照除去电辅助加热设备及病房热水器等三级负荷后，2 台变压器的合计计算容量及变压器过载能力，来确定变压器的负载率，为 60% ~ 75%，这也是变压器的最佳负荷运行点。值得注意的是，只在火灾发生时才投入使用的消防负荷不应计入计算容量。

为什么柴油发电机容量宜与变压器相同？因为这样做，柴油发电机可以完全承载 1 台变压器的容量，保证医院重要负荷的正常运行。同时，在变压器进线开关处切换，可以使系统简单可靠，操作与维修方便，无须过多的操作步骤，即可快速切换供电。另外，此容量的发电机组也是市场上的常见型号，库存较大，容易订货。

在部分箱式变电站订货时，要求其低压总开关更换为由 2 台框架断路器组成的 ATS（自动转换开关），便于发电机进线的接入和自动投切。发电机的启动信号由同组的 2 台变压器的高压侧 PT（电压互感器）提供，当 2 台变压器的高压侧电源均停电时，自动启动应急柴油发电机组。启动信号也可取自 2 台变压器的低压侧主开关之后的失压继电器。

室外箱式变电站及箱式发电机组设置于各区域的路边绿化带内，离污染区有一定的安全距离，便于操作维护，并采用景观式结构。其基础和接地根据厂家提供的大样图来实施。

1.5.5 低压配电系统

依据前文所述，除电辅助供暖、病房电热水器等三级负荷外，医院的所有负荷均须考虑 2 台变压器的备供。平时定制含 2 台变压器的箱式变电站是没问题的，但在非常时期几乎无法订货，而有现货的成品箱式变电站基本上只有单台变压器，实现 2 台变压器之间的低压联络较为困难。若增加联络柜，需要改造箱式变电站，会延长施工周期，同时室外的联络电缆需要多根并联，也会增加故障率。另外，2 台变压器之间的投切在设置机械联锁后只能采用手动方式也是不可接受的。因此，决定变压器之间不设置低压联络，这样，为实现重要负荷的 2 路供电电源的切换，必然得采用 2 路低压电源引入建筑物进行自动切换的方式供电。除电辅助供暖设备及病房电热水器设备等可以采用单路电源供电外，其余负荷均采用 2 路低压电源自动切换方式供电。消防负荷按照规范要求采用末端切换方式供电。

手术室、抢救室、监护室等 2 类医疗场所的供电应采用末端切换的方式供电，同时设置 UPS（不停电电源），并采用医用 IT 系统（医用隔离电源系统）。检验化验设备、血透设备及其他不允许停电的医疗设备、智能化系统设备除双电源末端切换外，均配置 UPS。UPS 的持续供电时间不小于 30min。这类特别重要负荷的供电线路采用由变电所直供的专线供电，不与其他负荷混合供电。

应急照明及疏散指示系统由自带蓄电池供电，供电时间按规范要求为 90min。

负压区域的通风设备、医用气体设备、放射设备、电梯设备均应采用专线供电，且采用 2 路电源末端自动切换方式。放射设备的供电线路应满足设备对电源内阻的要求。

除专线供电的设备外，其余设备按照分区配电方式供电，由变电所供电至建筑内进线间的进线配电箱，然后由进线箱配电至各区域配电箱，再由区域配电箱配电至末端配电箱。

建筑内的进线间、配电间及电气管井、弱电间及弱电管井、智能化系统机房均应设置于清洁区，并经清洁通道通向室外，以便于电气及智能化系统维修人员进场检修、维护和操作。消防及安防控制室应有直接对外的出口，而电气的总进线间、信息机房等也宜有直接对外的出口，因为电气操作及值班人员为非医疗人员，应尽量避开医疗场所，防止感染。

配电箱应设置在污染区外，这也是规范条文的规定，有人认为应设置在清洁区内，想法是对的，但难以全部做到，配电箱深入负荷中心的原则同样需要遵守。严禁配电箱设置于污染区，尽量设置于清洁区，是为了操作和维护的方便。进线配电箱及区域配电箱等主配电箱必须设置于单独的配电间及设备管井内，并置于清洁区，且通向室外的距离较短。

有些配电箱难以设置在清洁区，例如，病房配电箱需要设置于病房外缓冲间的走道侧门口，放射设备的配电箱设置于设备操作室内，这些都属于半污染区，属于例外。对于病房区这个重度污染区，除为单个病房服务的配电箱可设置于非污染区外，其他配电箱均不得在此区域设置。电气专业需要与建筑专业配合将电气设备间及管井设置于合适的清洁区内，同时配合暖通专业设置通风设备或空调，保证正常环境温度不高于30℃。

1.5.6 通风设备的控制

通风设备的控制要求应以暖通专业提出的要求为准，须满足以下要求。

通风空调系统的电加热器及电加湿器应与送风机联锁，并应设无风断电、超温断电保护及报警装置。送排风机发生故障时应能自动报警。负压隔离病房的排风机应与送风机联锁，排风机先于送风机开启，后于送风机关闭。风机同时应与密闭阀联锁。

由于未设置设备自动化管理系统（BAS），因此，空调通风设备最好由专业设备厂商配备控制器，按暖通专业的控制要求设置好逻辑控制关系，预先调试好再送至现场安装。电气专业应按照暖通专业的要求，对相关配电箱的回路留好控制接口并提出详细的控制要求。

空气压差控制（清洁区、半污染区、污染区之间的空气压差）是非常重要的环节。从可靠性方面来讲，风机采用定压模式是最好的。在电气专业的提议下，中南院暖通专业正是采用这种方式设计的，在送排风机选型及匹配时就通过计算压差确定了设备选型，无须电气专业采用自控方式来实现。如果暖通专业采用变频风机，然后由自控系统来实现压差控制，虽然暖通专业设计简单了，但为后期复杂的检测、控制和调试带来麻烦，影响工期。同时，一旦控制系统失灵，其后果也是不可想象的。雷神山医院的通风设计也是暖通专业和电气专业密切配合的范例，现场实测，其压差全部满足国家规范要求。

对于电气专业而言，只需在现场设置压差显示报警装置即可，具体做法为：负压隔离病房应设置监视病房与缓冲间、缓冲间与走廊之间压差的装置，安装于病房缓冲间外走廊侧的门口，当压差失调时应能声光报警；负压手术室、负压监护病房、负压检验室的压差报警装置的设置也同此，由净化工艺专业配合设计。

1.5.7 照明系统

照明系统设计应符合现行国家规范《建筑照明设计标准》（GB 50034—2013）、《医疗建筑电气设计规范》（JGJ 312—2013）及《传染病医院建筑设计规范》（GB 50849—2014）的要求。各场所的照度、功率密度、均匀度、显色性、眩光指数等照明指标应满足规范要求，不再赘述。

照明光源宜采用高效节能、显色性高的LED灯或T5型三基色荧光灯。病房、宿舍的光源宜采用暖色，色温不宜大于3300K，其余场所的光源宜采用中间色，色温宜为3300～5300K。LED灯的色温不宜大于4000K。LED灯蓝光危害指标不应高于RG1。

照明系统设计除传统的医疗照明设计内容外，需要注意的内容主要包括灯具选择、灭菌器的设置及应急灯的设置等。

医疗场所应选择不易积尘、易于擦拭的带封闭外罩的洁净灯具，不得采用格栅灯具。灯具采用吸顶安装，其安装缝隙应采取可靠措施密封。灯具安装位置应避开病床隔帘导轨等医疗设施。病房灯具安装应保证当隔帘拉上时，每床有一盏顶灯，便于医疗检查及治疗。病房及宿舍宜采用漫反射型灯具，避免卧床者产生眩光。病床设备带上应设置带线控的床头灯，便于患者阅读及休息。

病房内与病房走道设置夜间照明，宜由护士站统一控制。由于老年新冠肺炎患者较多，病房内的灯开关须兼顾老年人的使用，须采用宽板按键式开关，离地高度宜为1.2m。有人认为病房卫生间的灯开关应采用人体感应控制，其实没有这个必要，因为病房已是全面污染区，单纯地不触摸灯开关不会起到防污染的作用，况且病房内的主灯开关也是采用普通开关控制。感应灯在人体不动时会自动熄灭，反而会对患者不利。其他场所同理，也无设置人体感应开关的必要。病房照明、插座布置如图 1.64 所示。

走廊（有的规范写的是清洁走廊，其实不妥，因为病房走廊是半污染区域，不是清洁走廊，但同样需要设置消毒设施，而污染走廊在无人时也可能需要定时消毒）、病房、卫生间、更衣间、洗消间、候诊区、诊室、治疗室、手术室、各类医疗场所及其他需要灭菌消毒的场所须设置紫外杀菌灯或空气消毒器插座。紫外杀菌灯应采用专用开关，不得与普通灯开关并列，并有专用的警告标识及开启指示灯，距地高度宜为 1.8m。补充说明一下，紫外杀菌灯可破坏生物体的 DNA 或 RNA，对所有微生物均有杀灭作用，同样对人体有害，简单写上"紫外线灯"的标识起不到警告作用，所以应采用警告标识并带灯光指示，最好是可亮起的警告标识，如有告警声则更好，以保证仅在无人时才能开启。如采用人体探测方式进行联锁控制会更加安全，但实现起来有难度。

紫外杀菌灯开关应采用定时开关并带保护盒盖，消毒结束后自动关闭，消毒时间为 30 ~ 60min，可调整。开关应设置在消毒区外，便于人员撤离。一般来说，每立方米空间安装

图 1.64　病房照明、插座布置图

的紫外杀菌灯功率不小于 1.5W。灯具离地高度宜为 2 ~ 2.5m，至屋顶距离不宜大于 1.5m，距地面距离不宜大于 2.5m。

平时有人滞留的场所的紫外杀菌灯应在无人时使用，若需要在有人滞留时使用，应采用间接式紫外空气消毒器，紫外线密闭于消毒器内，不会伤害人体，通过主动吸气循环消毒，可以有效地对目标区域的空气进行消毒。需要注意的是，所谓的角度可调的紫外杀菌灯，并不是在有人时调节角度使其在不对准人后使用，而是在安装区域调节角度以使消毒无死角。任何直接型紫外杀菌灯都必须在无人时使用。

由于新型冠状病毒的传染性极强，几乎所有的场所都可能会用到灭菌设备，应按照紫外杀菌灯的保护范围设置足够的紫外杀菌灯，未设置的场所预留空气消毒器插座。

隔离病房传递窗口、感应门、感应便器、感应龙头、电动密闭阀等设施须预留电源。应与建筑和水暖专业配合确定位置。其电源须设置 30mA 漏电保护。

放射室、手术室、抢救室门上方设置工作警示标志灯。其电源接至照明回路，控制开关设置于门内侧，与门磁联锁。具体由净化公司设计安装。

应急疏散照明系统设计应符合《消防应急照明和疏散指示系统技术规范》（GB 51309—2018）的要求。

手术室、抢救室、重症监护室应设置安全照明，其照度值为一般照明的 100%。有的规范称为备用照明，其实不妥。根据 IEC（国际电工委员会）标准，对备用照明不作强制性要求，而对安全照明却有强制性要求，所以称其为安全照明较为妥当，由 UPS 保证供电，满足工作不中断的要求。

重要的医疗设备机房设置带电池应急灯，1类医疗场所每个房间至少有 1 个带电池的应急灯。电池备供时间不少于 30min。

1.5.8 线路选型与敷设

室内线路敷设时，普通负荷的电线电缆应采用低烟无卤阻燃型。消防负荷的电线电缆应采用防火型或低烟无卤阻燃耐火型。室外敷设应采用带聚氯乙烯护套的铠装电缆。

高压电缆由市政管网穿管埋地引至院区内的环网箱及箱式变电站。由变电站至各单体的低压电缆采用排管或电缆沟方式敷设。高压电缆与低压电缆应分开敷设，不宜共沟。排管经过道路时应采用混凝土包封。室外管线的敷设做法及与其他管线的间距应符合国家规范的要求，并应参照国家标准图集施工。应特别注意，室外管线敷设不能破坏室外地面的防渗膜，为便于检修，电气专业应与总图和给水排水专业配合，将防渗膜设置于排管或电缆沟的下方。另外，电缆电线引出屋面时，不应破坏屋面防水层，同时出屋面的电线管应采用防水弯头，防止雨水沿管孔倒灌进室内。同时管口应做好防水密封。这一点是设计人员容易忽视之处。

室内线路敷设时，主干电缆采用电缆线槽的方式敷设，沿顶板下吊装或墙壁安装，有吊顶处则在吊顶内安装。支线采用穿管敷设或小型线槽敷设，对于临时性医院的夹芯钢板型墙体，只能采用明敷方式。如果采用装配式，在工厂预制，可能会影响工期。若为预制混凝土墙体，则可以采用穿管开槽方式暗敷，也可采用明敷方式。水平线路可在吊顶内穿管敷设，或无吊顶时沿顶板面明敷。

线槽及穿线管应采用金属或无卤阻燃型 PE 材料，其开口处及穿墙的墙缝均应严密封堵。线槽及穿线管穿越污染区、半污染区及洁净区之间的界面时，隔墙缝隙及槽口、管口应采用不燃材料可靠密封，防止交叉感染。线槽穿墙处宜改为穿管敷设，便于封堵，否则，线槽内部难以做到可靠封堵。穿墙套管宜预留备用管。

对于必须进行射线防护的房间，其供电、通信电缆沟或电气管线严禁造成射线泄漏；其他电气管线不得进入和穿过射线防护房间。

2 类医疗场所局部 IT 系统的配电线缆宜采用具有电气绝缘的阻燃 PE 管敷设。

1.5.9 防雷及接地

应急医院的防雷、接地设计按现行的国家防雷及接地规范执行。

防雷类别宜按第二类防雷建筑物设计，临时性医院通常为钢结构建筑，其屋面为金属屋面，应与建筑专业配合，使其金属板厚度达到防雷规范的要求，可以直接作为接闪器。屋面天沟应敷设 25mm×4mm 镀锌扁钢或 $D12$ 镀锌圆钢作为避雷带，并与金属屋面连接。

防雷引下线可利用结构金属柱；接地装置可利用条形基础内的钢筋，当基础埋地深度不够时，应沿建筑物四周埋设人工接地带及接地极。具体应按照防雷、接地规范及标准图集的要求实施。接地电阻应实测不大于 1Ω。注意，建筑物外表面的金属墙板、金属屋面、防雷引下线及接地装置应连成电气整体。

室外箱式变电站及箱式柴油发电站应设置电源中性点接地及保护接地。具体做法以厂家大样图为准，并符合国家规范要求。接地电阻应不大于 4Ω。当距建筑物较近时，宜与建筑物接地系统连接。接地装置应设置于防渗膜的下面，不能破坏防渗膜的完整性。防渗膜若有局部破损应采取可靠的密封防水措施。

低压进线电源在入户后应实施重复接地，建筑物内采用 TN-S 系统。防雷接地、保护接地、功能性接地、屏蔽接地、防静电接地等共用接地系统。放射设备机房应设置屏蔽接地，放射室的金属设备、内墙金属屏蔽层、配电箱外壳、电源 PE 线均应与接地端子箱连接。医用气体管道应设置防静电接地，医用气体管道与支吊架接触处应做防静电腐蚀绝缘处理。

建筑物实施总等电位联结。ICU、手术室、抢救室、治疗室、淋浴间或有洗浴功能的卫生间等，应采取辅助局部等电位联结。

2类医疗场所内，设置医疗IT隔离电源系统及绝缘监测器、声光报警装置，该系统的等电位接地应采用单独接地方式。1类医疗场所、2类医疗场所中非用于维持生命的设备的回路采用额定剩余动作电流不超过30mA的剩余电流动作保护。

信息系统雷电防护，按A级保护类别设置SPD（浪涌保护器）。

消防控制室、安防监控室、信息机房等电子信息机房应设置机房接地系统及防静电地板，机房内所有设备的金属外壳、金属线槽、管道、防静电地板均须进行等电位联结接地，并与建筑物接地系统连接。

1.5.10 智能化系统

为便于快速设计及施工，智能化系统宜只选择必需的子系统，并且系统的形式以简单可靠为准。雷神山医院项目设置了如下系统：通信及计算机网络系统、综合布线系统、有线电视系统、安全防范系统、火灾自动报警及广播系统、呼叫信号系统等，并根据需要预留远程会诊系统。其他的智能化系统若无必要可不设置，以节省施工及调试时间。

其中，有线电视系统宜采用IPTV（交互式网络电视），并入综合布线系统；消防紧急广播与公共广播系统共用一套系统；FAS系统主要为火灾自动报警系统，按照有关消防规范来设计，无特殊要求。

移动电话信号覆盖系统由运营商及承包商负责同步设计与安装。医疗净化区（监护室、手术室、检验室）等场所的各系统点位须与工艺设计密切配合，满足工艺需求。

医院信息中心须预留与疾控中心、应急指挥中心及政府管理部门的通信接口。

智能化系统机房、消防控制室、监控机房等均应设置于清洁区，且有对外出口。

智能化系统的线槽及穿线管的口部应可靠密封，穿墙缝隙应严密封堵。

1. 通信、网络及综合布线系统

各大通信运营公司的通信光缆进线由城市通信管网穿管埋地引入医院通信信息机房，再由通信信息机房引至各单体电信间。室外通信线路采用9孔排管埋地敷设，经过道路时采用混凝土包封。

电话通信系统可采用虚拟交换方式，不设程控交换机。

计算机系统设置内网、外网和设备专网三套网络，且均采用物理隔离，并采用冗余的网络架构。内外网核心交换机采用跳线连接，实现信息互通。内网专门面向医护及管理人员；外网的使用对象为医护人员、行政人员及患者；设备专网用于承载各类基于TCP/IP协议的弱电通信信号，涵盖门禁及视频监控系统等。计算机网络采用含核心层、接入层的2层网络架构。

病房、护士站、办公室、会议室、医疗场所、宿舍等处按需求设置内网、外网、语音及IPTV信息点。语音及网络信息点的设置应满足国家卫健委和医疗机构的要求。信息点的设置应至少符合下列要求：

（1）病房的床头设备带上（2床共用）设置3个内网、1个外网信息点，病房内至缓冲间的门内侧设置1个电话语音信息点，病房内设置1个无线AP接入点（ICU须结合工艺设计预留网络点位，每床宜留4个内网信息点）。

（2）病房的病床对面墙上设置1个IPTV信息点（接外网）。

（3）护士站设置1个语音、3个内网信息点。

（4）医生、护士办公室每个工位设置1个语音、1个内网、1个外网信息点。

（5）处置室、治疗室、值班室、分诊台设置1个语音、1个内网信息点。

（6）诊断报告工作台、检验工作台、影像设备控制室、设备操作间、报告室、B超室、心电图室等每个工位设置1个语音、1个内网信息点（不含设备信息点）。

（7）医疗检验设备、检查设备（含超声、心电图等）、放射设备，每台设备设置1个内网信息点。

（8）医护宿舍每间房设置1个语音信息点、1个IPTV信息点（接外网）、1个无线AP接入点。

（9）会议室、会诊室设置1个语音、2个内网、2个外网信息点（须预留远程会诊系统的接入点）。

（10）候诊区的分诊台预留1个排队叫号系统信息点位（内网）。

无线网应实现院区全覆盖，接入点接于外网交换机，须同时生成2个SSID提供内外网服务（内外网采用逻辑隔离）。

特殊位置点位宜有冗余。ICU的无线信号覆盖密度须有保障。

网络机房应满足信息化系统所需的机柜安装条件，并配置UPS、精密空调、门禁和环境监控等。

污水处理站预留网络及电话接入点，便于水质在线监测，也可通过运营商的移动通信网络上传监测信息。

布线、监控基本安装调试完成的检验中心，如图1.65所示。

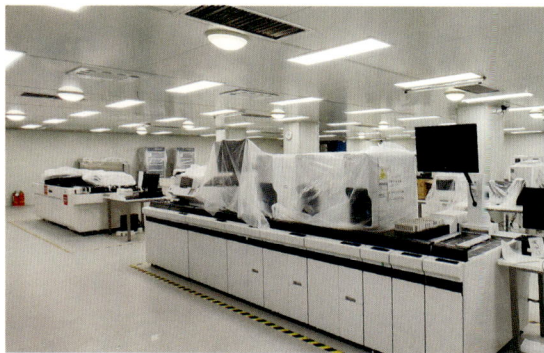

图1.65 布线、监控基本安装调试完成的检验中心

2. 安全防范系统

医院的安全防范系统包括视频监控系统、门禁系统、紧急报警系统、入口道闸系统等。

院区出入口及室外道路、楼栋出入口、各单体公共场所、候诊区、护士站、走道、防护衣更衣室、会议室、抢救室、手术室、ICU、血库、药库、配餐处、电梯轿厢（如有）等处设置视频监控系统，并采用数字视频综合管理平台。摄像机均采用低照度日夜型数字式高清网络摄像机。视频存储时间不少于30天。

病房区出入口、负压病房的医患通道、污染区与洁净区的过渡区设置门禁点，ICU及负压检验室缓冲间设置门禁点，并满足工艺上对A、B门联锁控制的要求。

门禁控制系统应根据医疗流程进行设置。对负压病房的医患通道、污染区与洁净区的过渡区宜进行控制，并应设置出入人员的识别功能，识别须采用非接触方式。门禁的污染区一侧宜设置紧急呼叫按钮，信号送至安防中心，以便门禁发生故障时医护人员可以呼叫打开。

当出现火灾等紧急情况时，根据火灾自动报警系统的联动信号，所有设置互锁功能的门都必须能处于可开启状态，所有处于疏散通道上的门禁自动打开。

护士站及医生办公室设置一键报警系统。监控室应有声光警报信号装置。隔离园区的车行入口设置车牌自动识别的道闸，人行入口设置带身份识别的人行道闸。

3. 呼叫信号系统

隔离医疗区的患者入口与接待室宜设置双向可视对讲系统，便于随时收治患者及安排病房，减少患者入院等候时间。

病区的各病房与护士站之间配置双向对讲呼叫系统，病房床头设置呼叫终端，卫生间设置呼叫按钮。

监护室、观察室等配置护士与病床之间的

双向对讲呼叫系统。

手术区配置护士站与各手术室之间的双向对讲呼叫系统。

放射科的控制室与放射设备室之间设置单向对讲系统。

ICU 设置远程移动探视系统，信号可由无线 AP 接入。其他病房则根据需要借用该探视系统，但通常需求量不大，不必考虑单设，因为一般患者可以通过手机利用无线网与家属可视对讲。

候诊区、放射科分诊登记台设置排队叫号系统。

1.6 空调通风设计

1.6.1 平面流线分析

医院隔离医疗区共分为 A、B、C 三个区。北区（A 区）拥有 15 个隔离病区、1 个药库区及位于东侧端头的医技区（含 ICU、检验中心、CT、超声波、心电图等功能区）。南区（B、C 区）拥有 15 个隔离病区、1 个药库区及位于东侧端头的 ICU，西侧医护通过单元与北区相连。病房区及医技区等患者到达的区域为污染区；医护人员经过更衣、消毒淋浴后进入的独立工作区为清洁区；清洁区与污染区之间的区域为半污染区，包括病区外的办公、会诊、治疗、护士站等用房。

依照"三区两通道"设计原则，患者流线设置于医院外围，严格与医护流线分开；医护流线设置于医院内部，保证医护人员不被感染。主要医护流线为：医护通过—清洁通道—清洁通过—隔离病区。每个单元区域在两端设置污物间，并设有单独的对外通道（图 1.66、图 1.67）。

1.6.2 设计总原则

1. 室内温湿度

为确保医务人员和患者身处良好的医疗环境，所有人员停留的房间均设有空调系统。空调设计室外参数主要考虑疫情暴发期间武汉市的冬季气象条件，主要房间空调设计温湿度参数见表 1.8。

2. 空调系统

空调系统划分及空调方式确定以平面区域划分及各区域功能为基础。为防止交叉感染，污染区、半污染区、清洁区均设置独立的空调系统。负压 ICU、负压检验室、负压手术室等房间采用直膨式全空气型净化空调机组，机组放置于专用的空调机房内。其他区域采用热泵型分体空调，室外机放置在室外地面或屋面上，采用壁挂式室内机。受货源不足限制，设置热泵型分体空调的区域的新风系统热源采用电加热器。

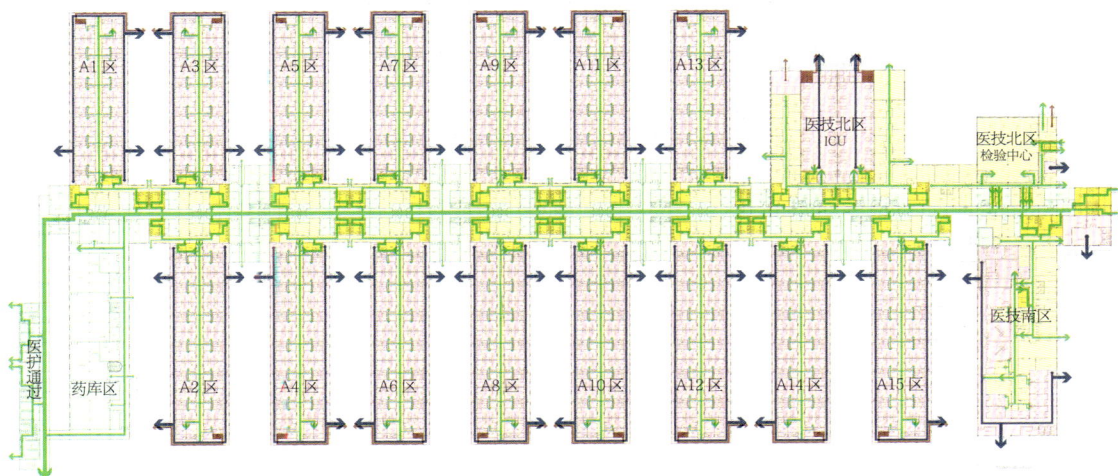

图 1.66 北区（A 区）内部流线图

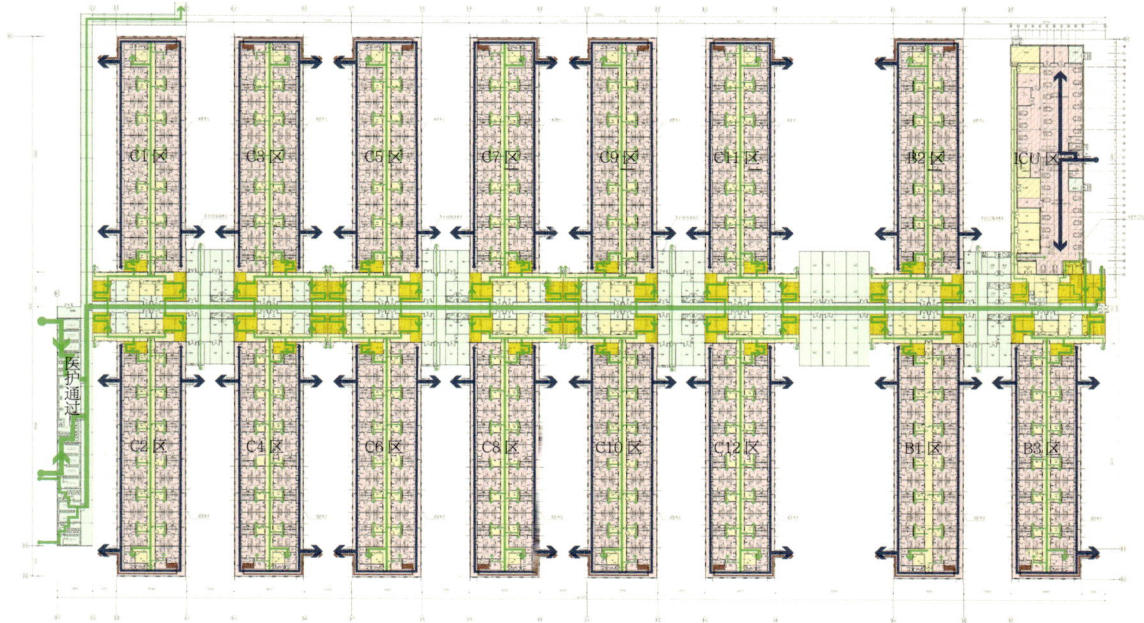

图 1.67　南区（B、C区）内部流线图

主要房间空调设计温湿度参数　　　　　　　　　　　　　　　　　　　　　　　　　　　　　　　　表 1.8

房间	温度 /℃	相对湿度 /%	房间	温度 /℃	相对湿度 /%
病房	18 ~ 22	—	负压 ICU	22 ~ 26	≤ 60
诊室	18 ~ 22	—	负压检验室	21 ~ 25	≤ 60
办公	20 ~ 22	—	负压手术室	21 ~ 25	30 ~ 60
医技检查室	18 ~ 22	—	负压手术室辅助用房及走道	21 ~ 26	30 ~ 60

电加热器具有分档（3 档）调节功能，且采取无风断电保护措施。空调冷凝水均随各区污水、废水分区集中收集处理。

3. 通风系统

通风系统应合理控制气流流向，保证有序的压力梯度，有效阻断病毒传播，保证医护人员安全健康。气流流动方向为清洁区→半污染区→污染区，相邻相通，不同污染等级房间的玉差不小于 5Pa。负压程度由高到低依次为病房卫生间、病房房间、缓冲前室与半污染走廊。清洁区气压相对于室外大气压保持正压。房间压差通过控制送排风量的差值形成，风量与压差的关系为 $Q=3600\mu F(\Delta p/\rho)^{1/2}$，式中 Q 为泄漏风量，单位为 m^3/h；μ 为流量系数，一般取 0.3 ~ 0.5；

F 为缝隙面积，单位为 m^2；Δp 为缝隙两侧空间压差，单位为 Pa；ρ 为空气密度，取 $1.2kg/m^3$。

空气中的悬浮颗粒物是病毒通过空气传播的主要载体，因此降低室内空气中的悬浮颗粒物浓度能有效阻止病毒传播。新型冠状病毒直径为 60 ~ 220nm，附着新型冠状病毒的悬浮颗粒物直径大于 $0.1\mu m$，H13 高效过滤器能有效过滤空气中直径为 $0.3\mu m$ 及以上的悬浮颗粒物，过滤效率高于 99.99%，因此，送风系统均设置 G2 粗效过滤器、F7 中效过滤器、H13 高效过滤器三级过滤器（图 1.68）。排风系统设置 H13 高效过滤器（医护清洁区除外），排风送至屋面（高 6.0m）排风口于高空处排放，新风取风口及排风口保持水平间距 20m，竖直间距 3.0m，

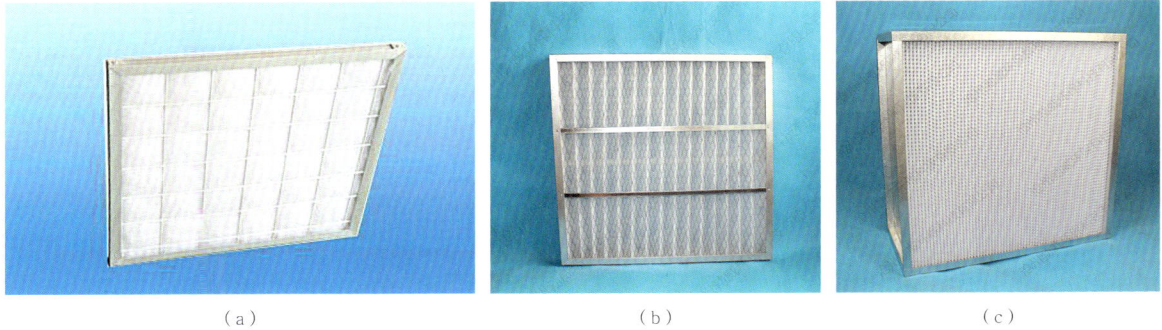

图 1.68　各种过滤器
（a）G2 粗效过滤器；（b）F7 中效过滤器；（c）H13 高效过滤器
来源：厂商产品宣传手册

图 1.69　部分厂商捐赠设备实景

避免送排风气流短路。通过采用上述过滤措施可有效保证送入空气的清洁度及安全性，同时避免排风对周边环境的污染。

4. 设备与材料选择

由于医院建设工期太紧，部分设备与材料的供应满足不了建设工期要求，如何因地制宜地根据现有的库存设备及材料来确定通风空调系统方案是设计师应重点考虑的问题。设计方案尽量选用成熟可靠、库存量大、运输快捷、厂商捐赠的设备（图 1.69），缩短产品的采购、调货时间，方便快速安装、简单调试，设备与材料的选择同时应满足国家有关规范及标准等的要求。例如，直膨式全空气型净化空调机组供应商无法在短时间内根据设计参数进行生产，在设计过程中设计师与供应商紧密对接，将现有的库存产品参数与设计值进行对比，第一时间锁定各地库存设备，并要求供货商对部分参数不达标的设备部件进行改造。

1.6.3　负压隔离病房

病房区主要用于收治重症确诊新冠肺炎患者，病房区采用双通道的平面布局形式，外围设置开敞式的患者走廊，中间设置半污染区的医护通道。病房入口及医护通道两侧均设置缓冲间。病房内设置卫生间，通道端头设置仪器室、污物间等功能房间。病房区平面图如图 1.70 所示。

1. 房间压力控制

根据病房区功能及工作流程，中部医护走廊为半污染区，病房为污染区。为确保各功能区之间的压力梯度满足工艺要求，负压隔离病房及半污染区医护走廊的机械送排风系统独立设置，并使空气压力从清洁区至污染区依次降低，污染区、半污染区保持负压。气流沿医护走廊→病房缓冲间→病房→卫生间方向流动，且相邻房间的压力梯度不小于 5Pa。负压隔离病房送风换气次数不小于 12 次 /h，排风量应在送风量的基础上增加维持房间压力值的渗透风量；医护走廊及缓冲区

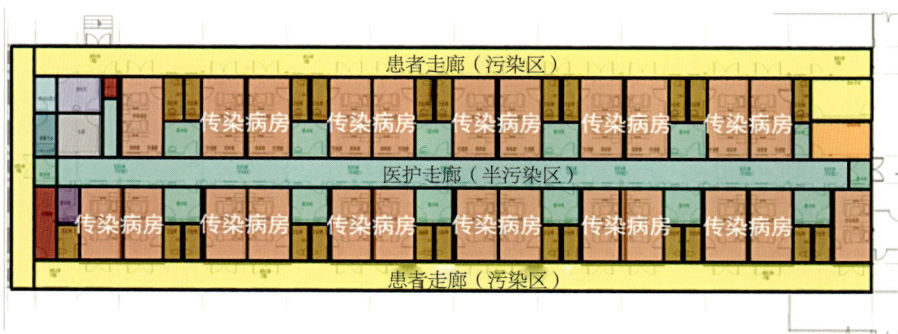

图 1.70　病房区平面图

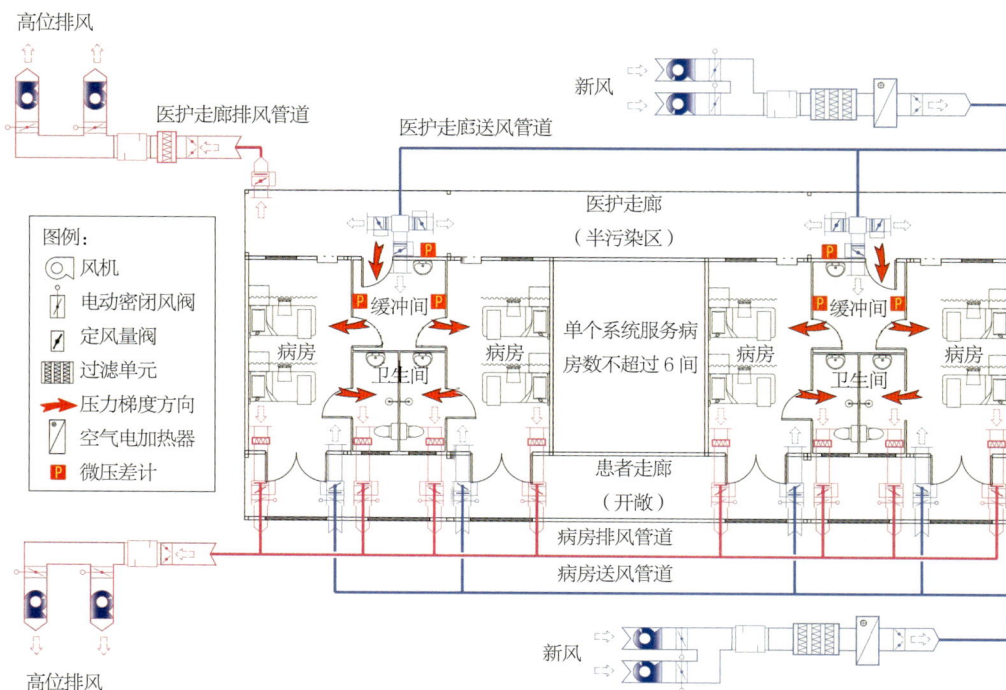

图 1.71　病房区送排风系统示意图

送风换气次数为 6 次 /h，走廊的排风量须根据维持压力值确定；缓冲间的排风量须根据其送风量及门缝隙渗透风量平衡确定，患者走廊为与室外空气相通的开敞走廊。病房区送排风系统示意如图 1.71 所示。

2. 管道与设备布置

病房区污染区、半污染区分别设置独立的送排风系统，每 6 间房间及其卫生间合用 1 套送排风系统。每间病房的送排风支管的定风量阀及电动密闭风阀设置在房间外走廊上部，既有效保证了压力梯度，又极大地方便了系统调试。

风机风量的合理控制减小了风机运行噪声和振动对病房中人员的影响。

负压隔离病房及其缓冲间的送排风口布置符合定向气流组织原则，缓冲间侧送风，病房顶部侧送风、下排风。经过粗效、中效、高效三级过滤处理后的清洁空气通过送风口送至病房医护人员停留区域，然后流过患者停留的区域进入排风口，保证气流流向的单向性，保障医护人员的安全健康。排风口设置在病房内靠近床头的下部并设高效过滤器，有利于污染空气就近尽快排出，且对周边大气环境不造成污染。送排风口如图 1.72 所示。

通风系统风机采用低噪声高效离心风机，考虑到新冠肺炎传染性强，所有通风系统均设置双风机（1用1备），提高系统运行的可靠性。风机及主风管设置在屋面，风机入口设置与风机联动的电动密闭风阀。病房及卫生间的送排风管均从侧墙直接进入室内，病房内未设置任何横向风管，空间简洁。医护走廊及缓冲间的送风管从医护走廊顶部进入后分别连接侧送风口，确保走廊净高合理，同时避免管道穿越污染区。所有送排风支管上均设置定风量阀，每间病房的送排风支管上均设置电动密闭风阀，可单独关断。送排风管如图1.73所示。

病房与医护走廊的墙面上装有显示不同区域压差的压差表（图1.74），便于医护和维护人员实时观察房间压力梯度并由此推断送排风系统运行是否正常。

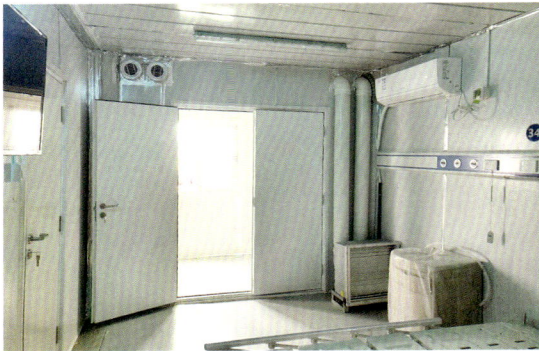

（a）

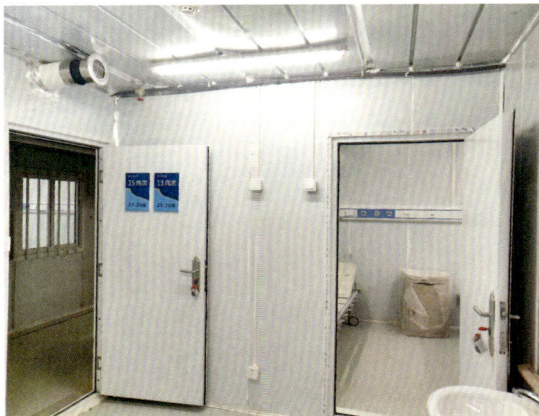

（b）

（c）

（a）

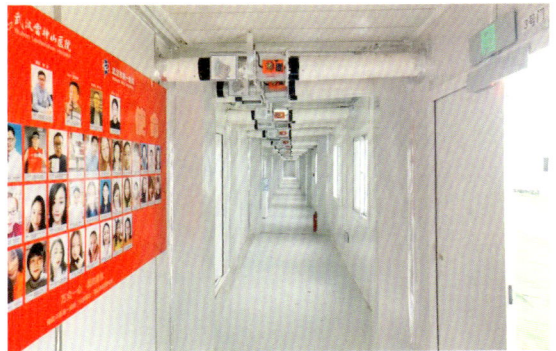

（b）

图1.72　送排风口的设置
（a）负压隔离病房送排风口实景；（b）缓冲间送风口实景

（d）

图1.73　送排风管的设置
（a）病房室外送排风管实景；（b）医护走廊室外送风管实景；
（c）病房室内送排风管实景；（d）医护走廊室内送风管实景

图 1.74 压差显示图

（a）病房与缓冲间压差显示；（b）缓冲间与医护走廊压差显示

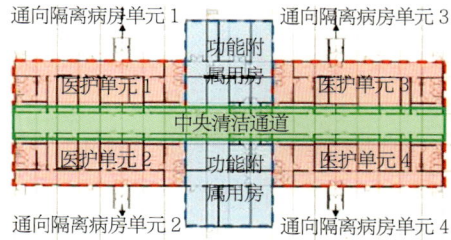

图 1.75 医护区平面图

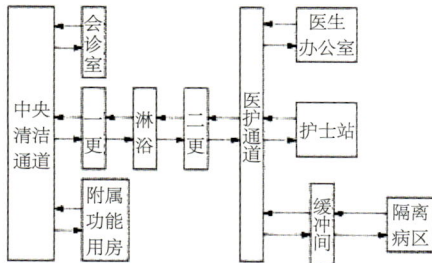

图 1.76 医护区工作流程图

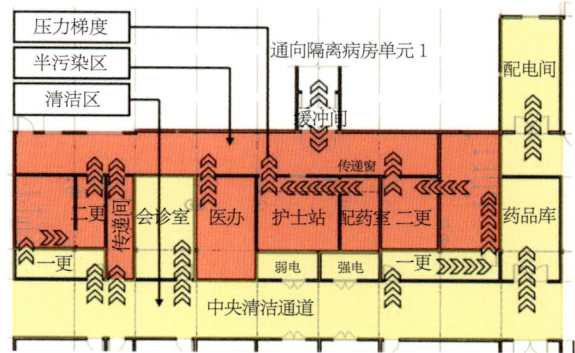

图 1.77 医护单元分区压力梯度示意图

1.6.4 医护区设计

医护区主要由中央清洁通道、医护单元、附属功能用房三部分组成（图 1.75）。中央清洁通道连接各个医护区及清洁用房（会诊室、休息厅及库房）；医护单元内设护士站、医生办公室等，通过医护通道及缓冲间与隔离病房单元连接；附属功能用房为清洁库房及医护休息室等。每个医护区含 4 个医护单元。

1. 房间压力控制

为确保各功能区之间的压力梯度满足工艺要求，且相邻房间压力梯度不小于 5Pa，玉力设定为：中央清洁通道 10Pa，一更 5Pa，二更、医护通道、医生办公室、护士站为 0Pa。为有效阻隔病房区污染空气流入医护区，医护区与病房区之间的缓冲间应保持正压，压力值为 5Pa，排气次数为 40 次 /h；清洁区送风换气次数为 4 次 /h，排风量为新风量的 50%；半污染区送排风换气次数为 6 次 /h。医护区工作流程如图 1.76 所示，医护单元分区压力梯度示意如图 1.77 所示。

2. 管道与设备布置

医护区的清洁区和潜在污染区的机械送排风系统均独立设置，具体设置如下。①清洁区：中央清洁通道、一更设置机械送风系统；会诊室、休息室设置机械送排风系统；清洁库房设置机械排风系统。②半污染区：二更、医护通道设置机械排风系统；医生办公室、护士站设置机械送排风系统。③医护通道与隔离病房单元间的缓冲间设置机械送风系统。医护区管道与设备布置实景如图 1.78 所示。

通风设备采用低噪声离心风机箱，风机及主风管设置在屋面上，风机入口设置与风机联动的电动密闭风阀。各房间送排风支管穿越屋面进入相关服务区域内，送风主管道上设置粗效、中效、高效三级过滤器，所有送排风支管上设置定风量阀。医护区通风系统示意如图 1.79 所示。

（a）　　　　　　　　　　（b）　　　　　　　　　　（c）

图 1.78　医护区管道与设备布置实景
（a）医护区房间送排风口；（b）医护区房间分体空调室内机；（c）医护区走道管线布置实景

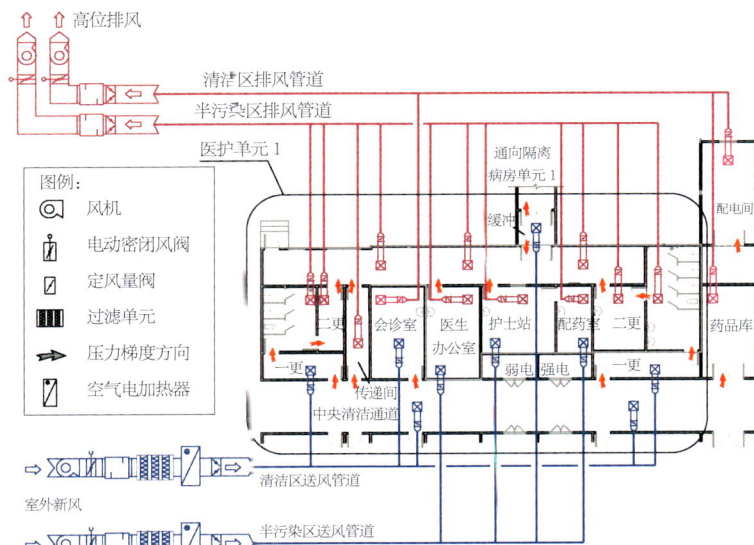

图 1.79　医护区通风系统示意图

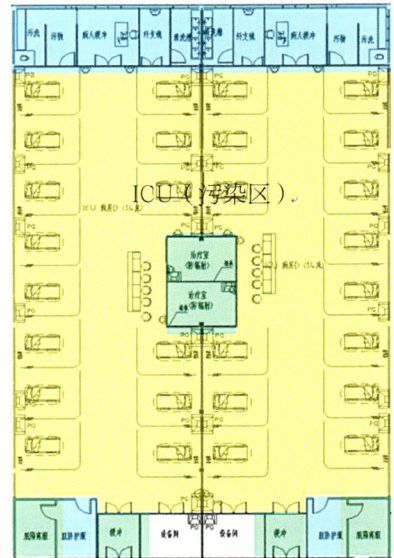

图 1.80　负压 ICU 平面示意图

1.6.5　医技区设计

医技区主要包括负压 ICU、负压检验中心、负压手术室。医技区分为 A、B、C 三个区，A 区主要包括负压 ICU、负压检验中心；B 区包括负压手术室；C 区为负压 ICU。

1. 负压 ICU

医技区 A、C 区分设负压 ICU（图 1.80）。负压 ICU 主要由病房区　患者缓冲、污物、污洗、清洗槽、纤支镜、脱防护服、脱隔离服、治疗室、缓冲等部分组成。负压 ICU 设独立的机械送排风系统（图 1.81）。气流须沿脱隔离服、治疗室、缓冲→患者缓冲、污物、污洗、清洗槽、纤支镜、脱防护服→病房区方向流动，且相邻房间压力梯度不小于 5Pa。负压 ICU 主要房间换气次数及负压值见表 1.9。

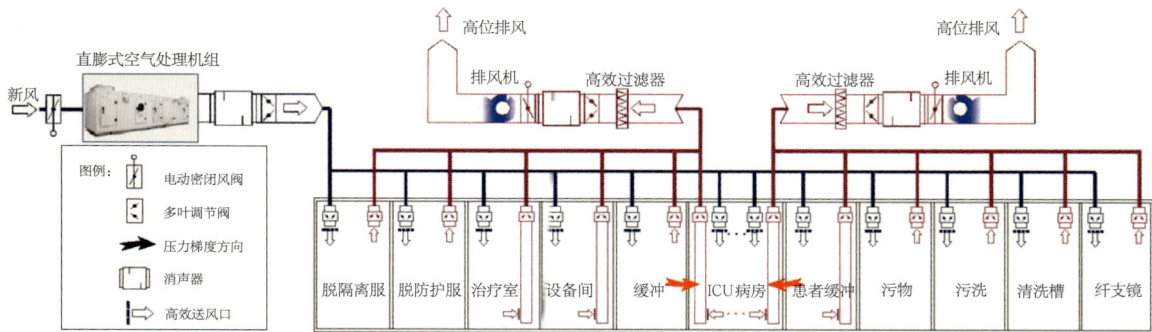

图 1.81　负压 ICU 通风系统示意图

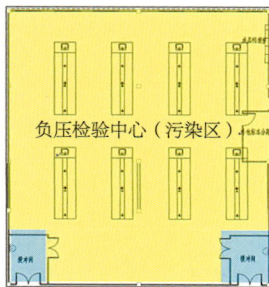

图 1.82　负压检验中心平面图　　图 1.83　负压检验中心·通风系统示意图

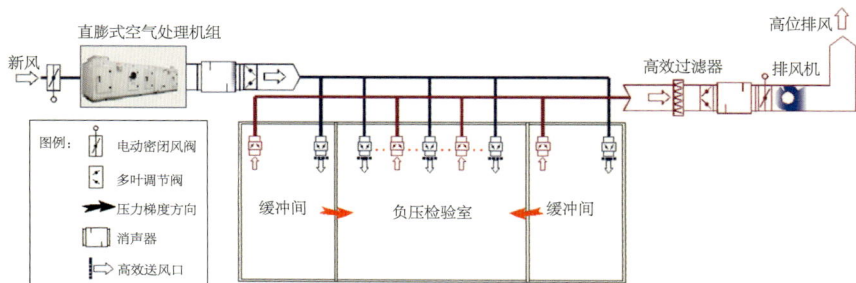

负压 ICU 主要房间换气次数及负压值　　　　表 1.9

房间功能	换气次数 /（次 /h）	负压值 /Pa
病房区	13	−20
患者缓冲、污物、污洗、清洗槽、纤支镜、脱防护服	13	−15
脱隔离服、治疗室、缓冲	13	−10

2. 负压检验中心

医技区 A 区设负压检验中心（图 1.82）。主要由负压检验室及缓冲间组成。2020 年 1 月 23 日，国家卫生健康委员会办公厅发布了《新型冠状病毒实验室生物安全指南（第二版）》，指出新冠病毒病原体暂时按照病原微生物危害程度分类中第二类病原微生物进行管理。新冠病毒具有气溶胶传播的可能，新冠病毒病原体检测应至少在生物安全二级实验室中进行，最好在加强型二级生物实验室中进行。负压检验室空调通风系统主要参考《生物安全实验室建筑技术规范》（GB 50346—2011）对二级生物安全实验室的相关要求进行设计。

负压检验中心设独立的机械送排风系统，整体维持负压（图 1.83）。气流沿缓冲间向负压检验室方向流动，且相邻房间压力梯度不小于 10Pa。负压检验中心主要房间换气次数及负压值见表 1.10。

负压检验中心主要房间换气次数及负压值　　　　表 1.10

房间功能	换气次数 /（次 /h）	负压值 /Pa
负压检验室	13	−20
缓冲间	13	−10

3. 负压手术室

医技区 B 区设负压手术室 1 间，主要由手术室及辅助用房两部分组成。辅助用房包括前室、消毒打包、复苏、存床、走廊等空间或房间。负压手术室平面布置如图 1.84 所示。负压手术室设独立的机械送排风系统，整体维持负压（图 1.85）。气流沿走廊、存床、医护前室、无菌间→复苏室、消毒打包、前室→手术室方向流动，且相邻房间压力梯度不小于 5Pa。负压手术室主要房间换气次数及负压值见表 1.11。

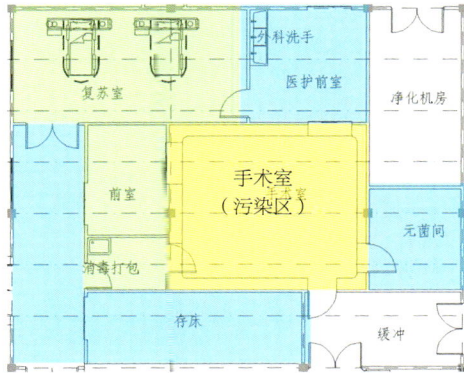

图 1.84　负压手术室平面示布置

负压手术室主要房间换气次数及负压值　　　　表 1.11

房间功能	换气次数 /（次 /h）	负压值 /Pa
手术室	18	-20
复苏室、消毒打包、前室	12	-15
走廊、存床、医护前室、无菌间	12	-10

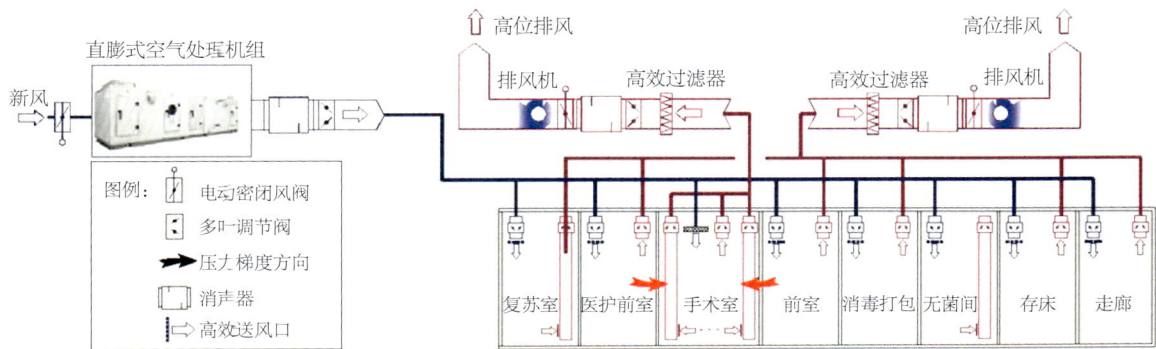

图 1.85　负压手术室通风系统示意图

4. 管道与设备布置

送风设备采用直膨式全空气型净化空调机组，排风设备采用低噪声离心风机箱，均设置在专用空调机房及风机房内。负压 ICU、负压手术室气流组织为上送下排，排风口的底部设在房间地板上方不低于 100mm 的位置。清洁空气通过送风口送至医疗人员停留区域，然后流过患者停留的区域进入排风口，保证气流流向的单向性。负压检验中心气流组织为上送上排，保证负压检验中心送排风的整体均匀性。

送风系统设置粗效、中效、高效三级过滤器，排风经高效过滤器处理后于高空处排放。同时为确保排风管路内的污染空气不外溢污染其他功能房间，排风管内压力设计为负压，风机设置在排风管路的末端，排风机吸入口设置与风机联动的电动密闭风阀。医技区建成实景如图 1.86 所示。

1.6.6　检查区设计

检查区主要位于医技区 B 区，包括 CT 室、DR 室、心电图室、医生办公室、休息室等房间（图 1.87）。

1. 房间压力控制

检查区医生办公室、休息室等清洁区换气次数：送风 6 次 /h、排风 5 次 /h。医护通道等半污染区换气次数：送风 6 次 /h、排风 6 次 /h。各检查室及配套用房污染区换气次数：送风 6 次 /h、排风 8 次 /h。通过送排风系统换气次数的差异确保各功能区之间的压力梯度，确保气流沿清洁区→半污染区→污染区方向流动。污染区、半污染区、清洁区之间保持不小于 5Pa 的负压差。

2. 管道与设备布置

医技区中检查的清洁区、污染区、半污染区的机械送排风系统均独立设置（图 1.88），设

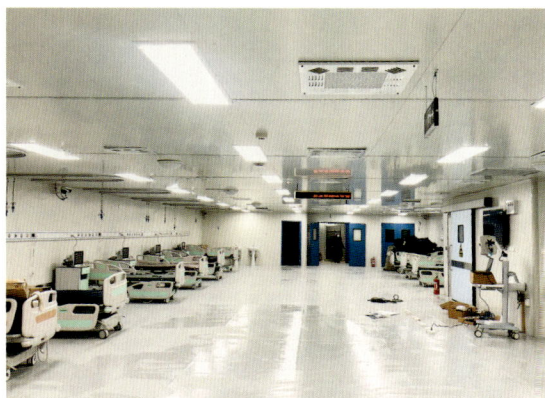

(a)

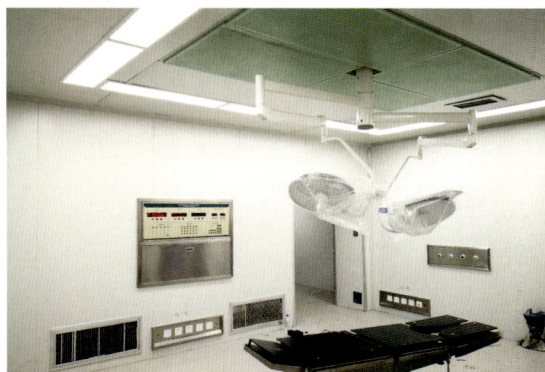

(b)

(c)

(d)

图 1.86　医技区建成实景图
（a）负压ICU；（b）负压手术室；（c）负压检验中心；（d）医技区

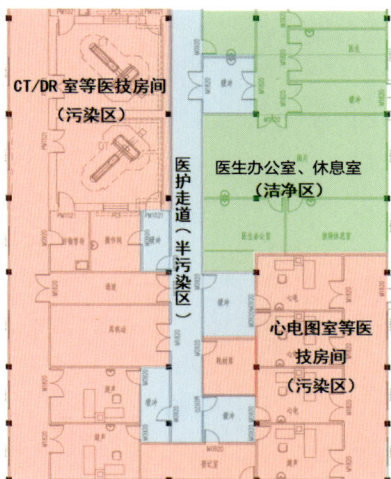

图 1.87　检查区平面示意图

备采用低噪声离心风机箱，风机放置在专用风机房内。按照压差需求计算确定房间送排风风量，各房间送排风支管上均设置定风量阀。所有机械送风系统均设置粗效、中效、高效三级过滤器，排风通过高效过滤器过滤后在高空处排放。

CT室、DR室实景如图 1.89 所示。

1.6.7　病房气流组织模拟

护理单元建筑为集装箱拼接式建筑，外形尺寸为3m（宽）×6m（长）×2.9m（高），室内净高2.4m。为了确保室内空间简洁，设计时应尽量避免在病房内设置横向风管，因此，负压隔离病房采用了同侧上送下排的气流组织方式。为了研究病房内气流组织及含病毒污染物的流动情况，在设计后期，中南院工程数字技术中心联合法国达索系统公司进行了气流组织和污染物浓度模拟，基于 XFlow 软件对比分析了不同送排风方案对室内气流及污染物浓度的影响，并根据模拟结果确定了医护人员相对安全的活动区域（图 1.90）。模拟结果表明，上侧送风和下侧排风的气流组织可以在病床处形成回流区，及时有效地排除病房内的污染空气。建议医护人员的主要活动区域集中在靠近缓冲间、观察窗和传递窗一侧，避免靠近送风口处及卫生间。

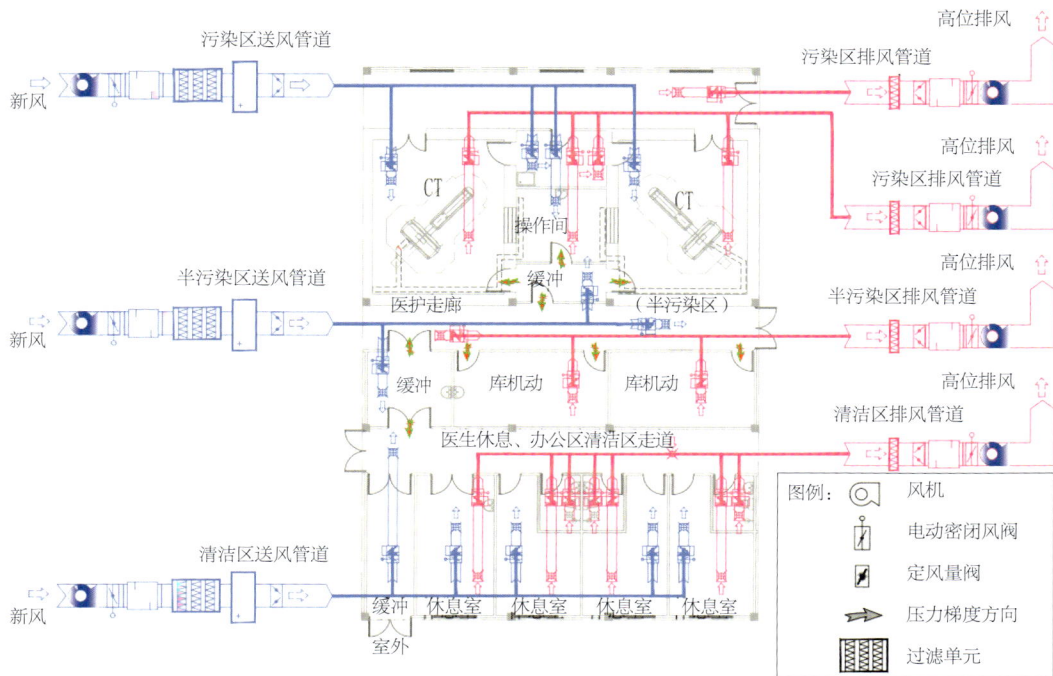

新风　污染区送风管道　　　　高位排风
污染区排风管道
新风　半污染区送风管道　　　　高位排风
污染区排风管道
医护走廊　（半污染区）　高位排风
半污染区排风管道
缓冲　库机动　库机动　高位排风
清洁区排风管道
医生休息、办公区清洁区走道
新风　清洁区送风管道
缓冲　休息室　休息室　休息室　休息室
室外

图例：　风机
电动密闭风阀
定风量阀
压力梯度方向
过滤单元

图 1.88　检查区通风系统示意图

（a）　　　　　　　　　　　　（b）

图 1.89　CT室、DR室实景图
（a）CT室；（b）DR室

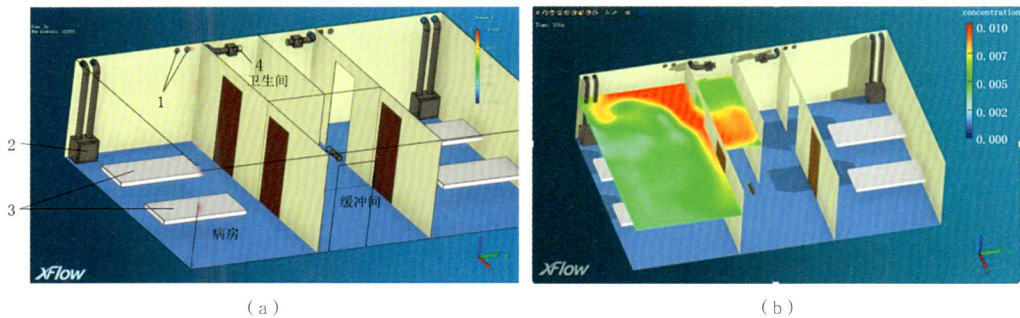

（a）　　　　　　　　　　　　（b）

1—送风口；2—病房排风口；3—床位；4—卫生间排风口
图 1.90　三维模拟图
（a）三维模型示意图；（b）医生站立高度（Z=1.5m）与污染物浓度对比
来源：法国达索系统公司　提供

1.6.8 病房室内温度预判及应对策略

　　新冠肺炎疫情暴发正值冬季的农历新年，设备货源不足，为使病房温度保持在18～22℃，每间病房设置了热泵型分体空调，同时新风系统的热源采用电加热器。在考虑冬季制热方案的同时，设计团队对病房在制热季结束后即将到来的制冷季的室内温度进行了预判，并提出了相应的应对措施。在制冷季病房室内温度不宜超过26℃，武汉市6月的室外干球温度为32℃，热泵型分体空调的实际制冷量须进行相应修正。设计团队采用DeST软件对4月、5月、6月，病房室内温度分别为24℃、25℃、26℃，送风量（含房间渗透风量）分别为400m³/h、550m³/h、800m³/h时的逐时冷负荷进行了计算，并将其与分体空调实际制冷量进行对比，统计出不满足对应设计温度的时长，详见图1.91。

　　同时根据4月、5月、6月最高温当天逐时室外干球温度，统计出室内温度分别在24℃、

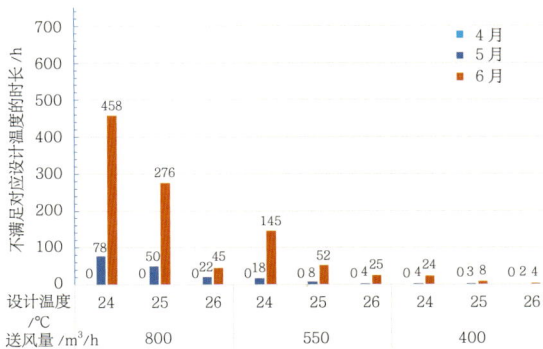

图1.91　4月、5月、6月病房不满足对应设计温度的时长
注：4月总时长为720h，5月总时长为744h，6月总时长为720h。

25℃、26℃，送风量在400m³/h、550m³/h、800m³/h时的逐时冷负荷与分体空调制冷量对比，如图1.92～图1.94所示。

　　对以上数据进行分析可以对病房室内温度进行预判并提出应对策略：为使病房室内温度维持在24～26℃，4月可不调节房间送风量；5月须调节房间送风量；6月可根据室外温度增加送风系统制冷设备。

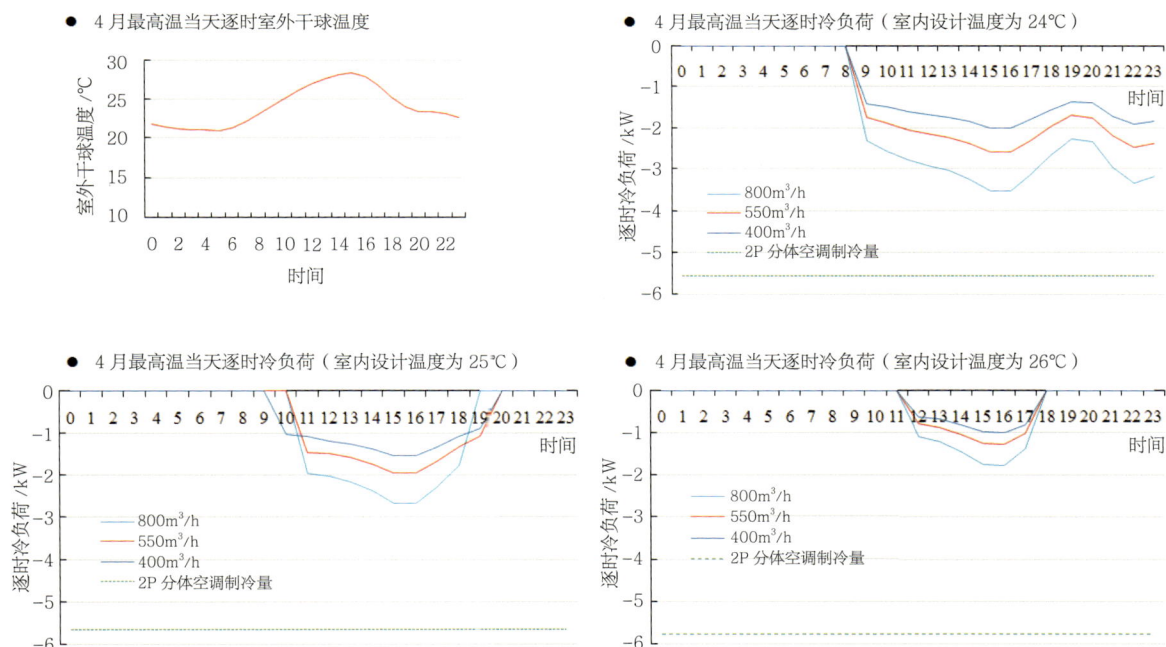

● 4月最高温当天逐时室外干球温度

● 4月最高温当天逐时冷负荷（室内设计温度为24℃）

● 4月最高温当天逐时冷负荷（室内设计温度为25℃）

● 4月最高温当天逐时冷负荷（室内设计温度为26℃）

图1.92　4月最高温当天逐时冷负荷与分体空调制冷量对比

● 5 月最高温当天逐时室外干球温度

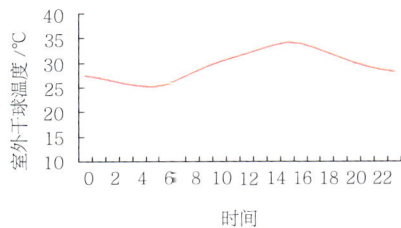

● 5 月最高温当天逐时冷负荷（室内设计温度为 24℃）

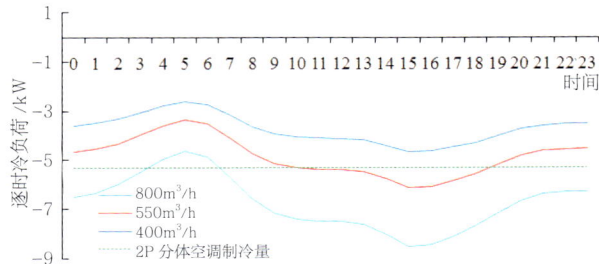

● 5 月最高温当天逐时冷负荷（室内设计温度为 25℃）

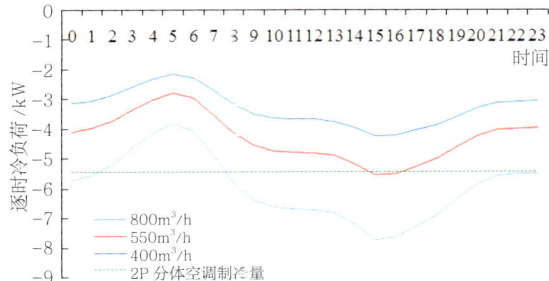

● 5 月最高温当天逐时冷负荷（室内设计温度为 26℃）

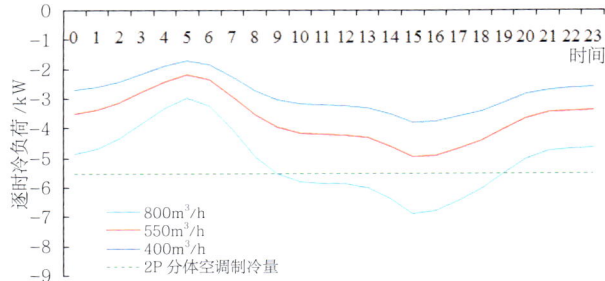

图 1.93 5 月最高温当天逐时冷负荷与分体空调制冷量对比

● 6 月最高温当天逐时室外干球温度

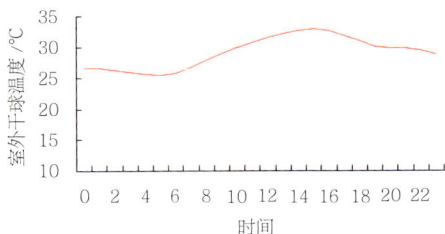

● 6 月最高温当天逐时冷负荷（室内设计温度为 24℃）

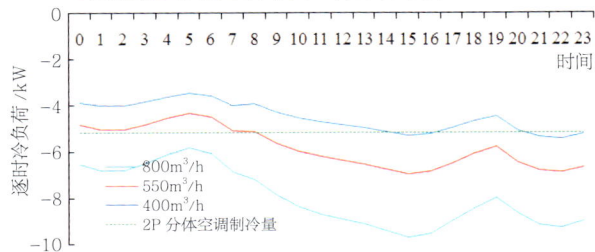

● 6 月最高温当天逐时冷负荷（室内设计温度为 25℃）

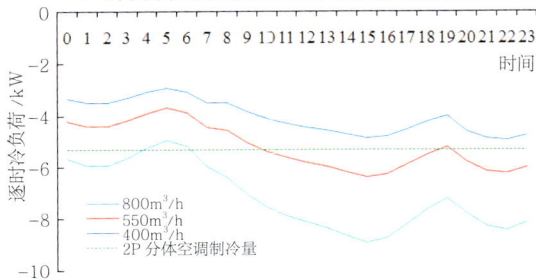

● 6 月最高温当天逐时冷负荷（室内设计温度为 26℃）

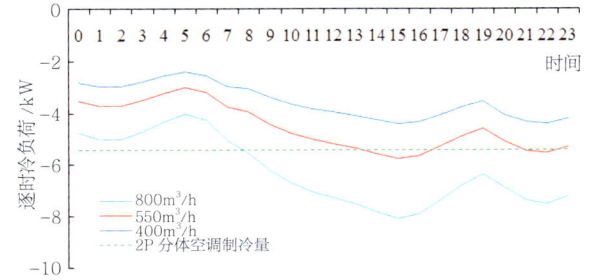

图 1.94 6 月最高温当天逐时冷负荷与分体空调制冷量对比

1.6.9　污染空气排放数值模拟

1. 模型建立及分析

为充分评估污染空气排放对项目周围环境造成的影响，项目组特邀清华大学陆新征教授及其团队进行模拟分析。陆新征教授及其团队提出了一种可快速模拟临时医院排风环境影响的方法。该方法以开源流体力学计算软件FDS为基础，实现了临时医院建筑的快速建模、基于云计算平台的分布式计算及污染空气流动的监测和可视化，为临时医院设计阶段的快速分析提供了依据。污染空气在空中运动的示踪粒子轨迹和浓度等值面模拟结果如图1.95所示，其中黑色粒子为污染空气的示踪粒子；红色曲面为污染空气的浓度等值面。

如图1.96所示为模拟得到的不同排风口高度下新风口高程处的污染空气相对浓度，可见，将排风口高度提高后，距室外地面3.0m处的污染空气最大浓度有了显著降低。

2. 主要结论与指导

（1）将排风口距室外地面高度从4.5m提高到6.0m后，可显著降低新风口高程处（距室外地面3.0m）的污染空气相对浓度。

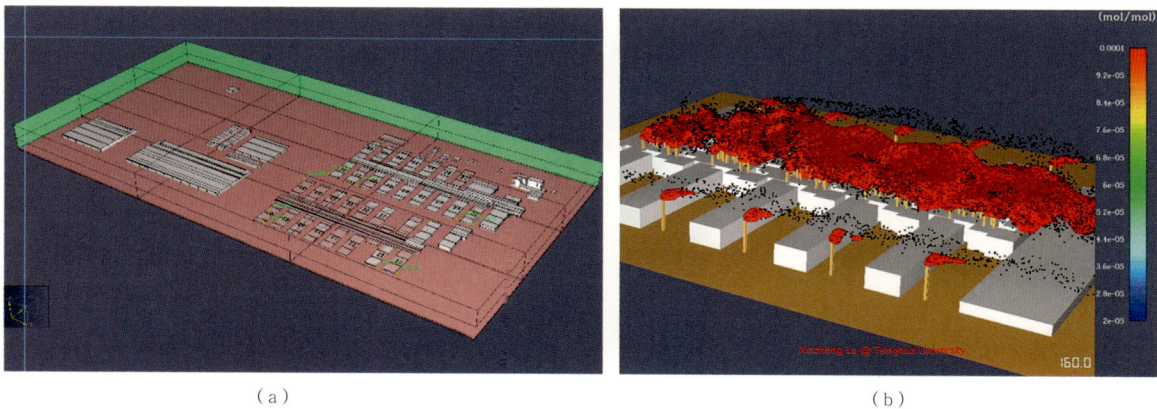

（a）　　　　　　　　　　　　　　　（b）

图1.95　污染空气在空中运动的示踪粒子轨迹和浓度等值面模拟结果
（a）雷神山医院三维FDS模型；（b）污染空气轨迹及浓度等值面图
来源：清华大学陆新征　提供

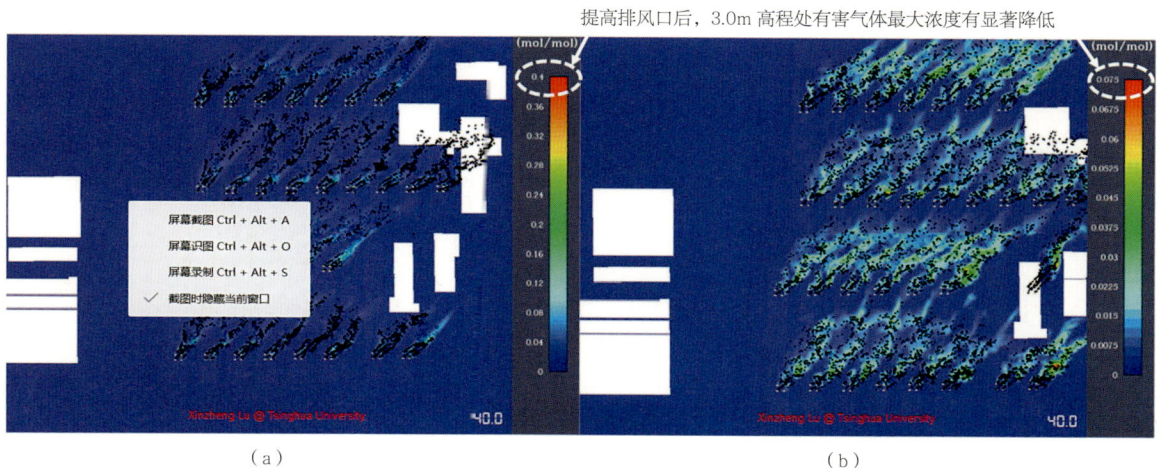

（a）　　　　　　　　　　　　　　　（b）

图1.96　不同排风口高度下新风口高程处的污染空气相对浓度
（a）排风口距室外地面高度4.5m；（b）排风口距室外地面高度6.0m
来源：清华大学陆新征　提供

（2）提高排风口高度后，新风口高程处的污染空气最大相对浓度可从 2.6%（相当于稀释 3.8 万倍）降低至 2.0%（相当于稀释 5.0 万倍）。

（3）根据江亿[29]等人的研究，SARS 病毒稀释 1 万倍后不再具备传播性。排风口设在距室外地面 4.5m 和 6.0m 的高度都可以满足新风稀释 1 万倍的要求。

（4）为确保负压隔离病房区域医护人员的健康安全及保护室外环境，该项目送风经粗效过滤器（G2）、中效过滤器（F7）、高效过滤器（H13）三级过滤，排风经高效过滤器（H13）过滤后接至屋面（距室外地面高度 6.0m）在高空处排放（图 1.97）。通过该模拟进一步论证了设计方案的合理性。

1.6.10 室内环境监测

清华大学林波荣教授团队疫情期间在雷神山医院典型区域布置了 44 套无人值守的云端在线分布式室内环境监测设备，对医院患者和医护人员停留区域的室内 CO_2 浓度、$PM_{2.5}$ 浓度、温度、相对湿度等参数进行高密度、高精度的实时监测。图 1.98 所示为环境监测设备布置区域示意图。

环境监测设备采用智能建筑室内环境监测仪，该仪器集成了温度、相对湿度、$PM_{2.5}$、CO_2 等传感器，可实现长期在线监测。同时在手机端搭建医院室内环境实时在线监测系统平台（图 1.99），读取监测仪采集的实时数据。工作人员可以通过手机等设备读取医院室内环境的各项参数，进而了解医院内部环境控制设备的运转情况，当监测参数出现异常时，平台可以发出预警。

从典型病房环境参数 24h 变化图（图 1.100）可以看出，CO_2 浓度在 520～670ppm 范围内波动，在 14：30 及 21：00 出现峰值，这是由医护人员进病房查房，人员密度增大导致的。$PM_{2.5}$

图 1.97　排风风帽现场安装高度实景

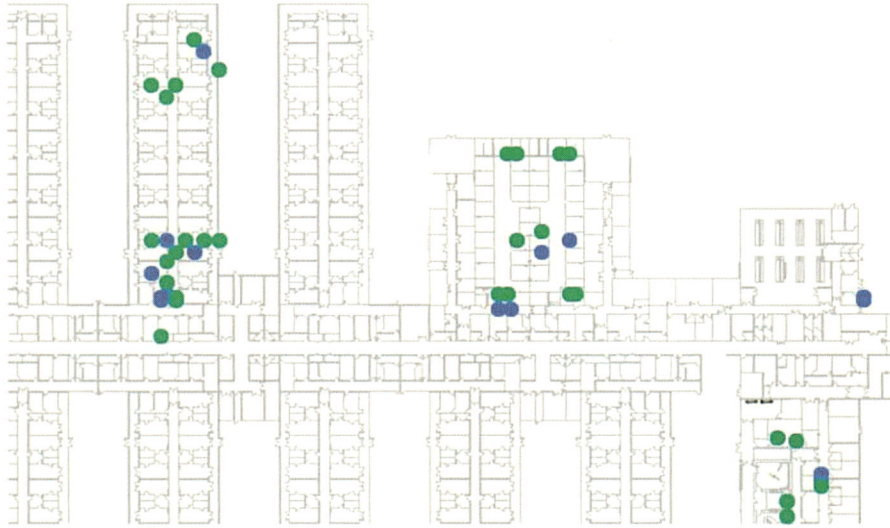

图 1.98　环境监测设备布置区域示意图

来源：清华大学林波荣　提供

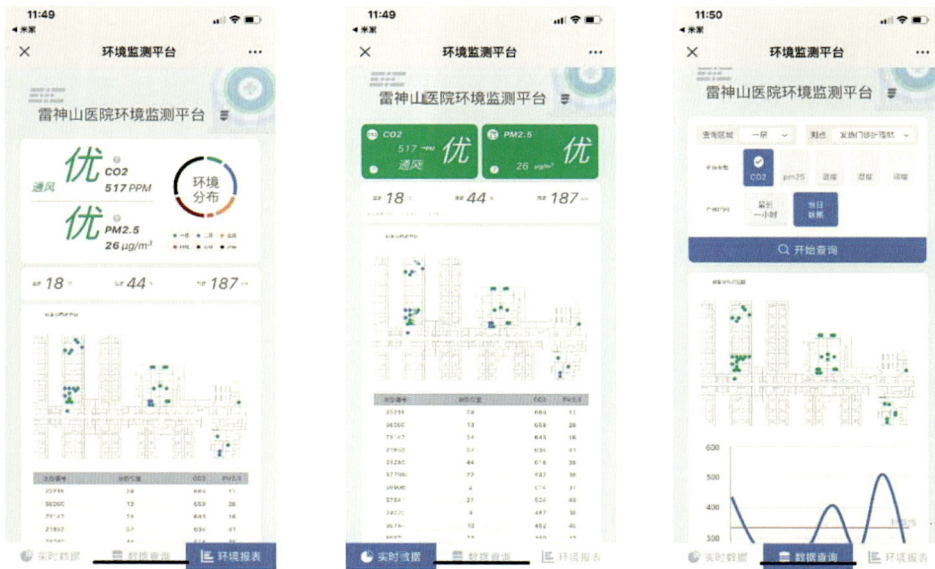

图 1.99　雷神山医院室内环境实时在线监测系统平台

来源：清华大学林波荣　提供

浓度在 11：00 上升至顶峰后于 14：00 回落，说明此时污染物浓度较大，防护压力较大。其余时间比较稳定，浓度线趋于直线。房间温度在 20 ~ 22℃范围内波动，相对湿度在 40% ~ 55% 范围内波动。主要时间段内的监测数据均满足设计要求。

1.6.11　安装与调试

1. 安装应重点关注的问题

（1）由于临时医院建设周期极短，多专业、多工种交叉施工，因此施工时不同工种间的相互配合十分重要。

（2）通风空调安装技术人员应及时充分熟

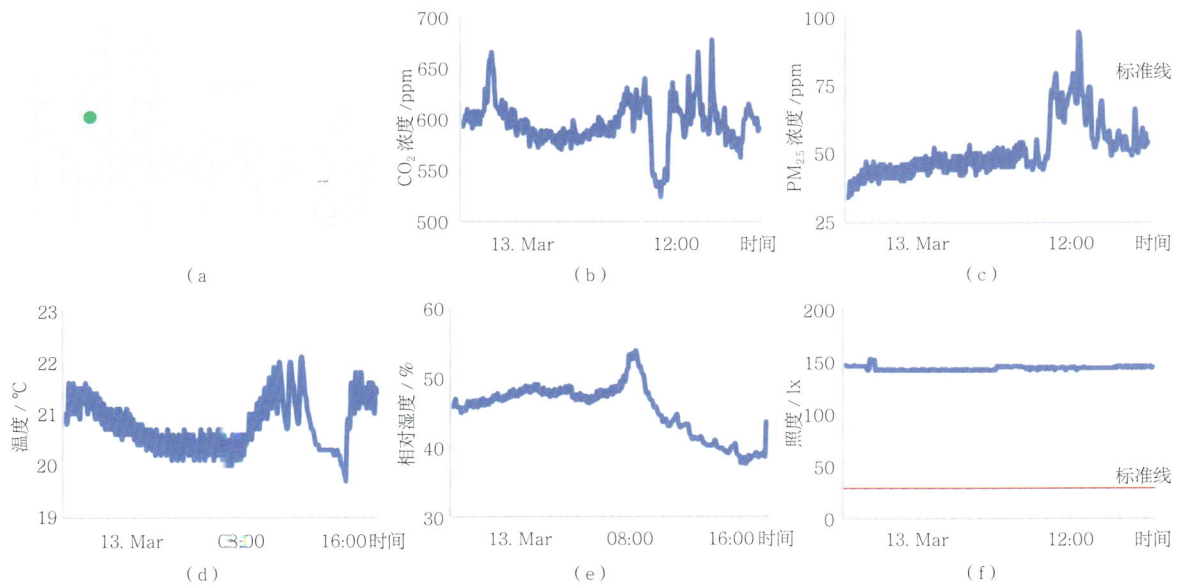

图 1.100　典型病房环境参数 24h 变化图

（a）设备分布点位；（b）CO_2 浓度监测；（c）$PM_{2.5}$ 监测；（d）温度监测；（e）相对湿度监测；（f）照度监测

注：CO_2、$PM_{2.5}$ 浓度数值采用监测仪中给出的数值单位 ppm 计量。

来源：清华大学林波荣　提供

悉设计图纸，同时结合现场实际情况尽快制定详细周密的施工方案，包括管道加工、制作、安装方案，风管严密性检测方案，系统调试运行方案等。

（3）由于护理单元的建筑为室内净高 2.4m 的集装箱拼接式建筑，工程安装过程中要特别注意走道及房间内的机电管线综合排布，避免管线布置影响室内净高。

（4）确保风管、风阀、风机之间接管的严密性。确保围护结构的气密性，特别注意风管、冷媒管等管道穿越屋面及侧墙等围护结构时的防水及密封处理，病房拼接处的缝隙也须认真处理，确保不留死角，为维持合理的室内压力梯度创造先决条件。

（5）每间病房送排风支管上的定风量阀、电动密闭风阀均安装在病房外侧敞开走廊上方，便于系统调试及后期维修管理。

（6）每间病房下部排风口处的高效过滤器在设计中应留出足够的安装及更换空间，便于

后期更换。

（7）由于箱式房构造无法承受设备及管道荷载，因此应在箱式房的 4 个立柱上设置槽钢作为风机、过滤单元、电加热器及风管等设备的承重构件。

（8）为满足建设工期的要求，室外风管由 PE 管代替常规风管，施工快捷，气密性好。

2. 调试要点

（1）病房通风系统调试尤为重要，因为工期紧，故要求尽可能一次调试成功。

（2）通风系统调试以相邻房间的压差满足设计及使用要求为目标。各房间均须进行风量平衡调试，调试时各房间内的门窗必须紧闭。

（3）系统调试时，定风量阀的风量按照设计文件中规定的风量进行设定。

（4）为保证各区域间的压差合理，无特殊情况，不得对已设定的定风量阀进行二次调整。如遇特殊情况，须记录下各阀门的开启角度，后期按照同样的开启角度进行复原。

（5）应重点关注负压隔离病房、负压ICU、负压检验室、负压手术室等场所的压力梯度，保证各区域房间的压力梯度满足设计及使用要求。

3. 调试结果

经过各系统的调试，各区域通风及空调系统运行正常，病房区、医护区、医技区的温湿度、换气次数、压力梯度等满足设计及使用要求。

1.6.12　总结与思考

及时总结应急医院的设计建设经验，对控制病毒的传播、提高应急医院的可靠性、有效救治患者及高效保护医护人员有着重要意义。本节详细解读了雷神山医院通风空调设计理念，认为应急医院不仅应满足常规医院的基本功能要求，而且还应针对应急医院建设工期短、设备材料货源短缺、建筑空间不足、医院为临时医疗建筑等特点，快速高效地组织设计。主要总结与思考如下。

（1）"三区两通道"是应急医院的重要设计原则，在"三区两通道"区域控制合理的压力梯度，确保气流的定向流动是通风设计的关键。

（2）应急医院建设工期短，为满足快速安装的要求，空调系统形式选择应因材制宜，空调通风系统设备选择必须考虑有现成的货源，确保医院建设工期及空调通风效果是首要考虑因素。

（3）因工期原因，不建议在系统中设计复杂的自动控制系统，建议用简单实用的方式达到使用要求。雷神山医院项目设计中采用定风量阀固定各区域送风量及排风量，简单有效地保证了各区域的压力梯度，满足了使用要求。

（4）由于护理单元的建筑为集装箱拼接式建筑，在压力控制中应高度关注房间气密性，同时也应关注管道穿越围护结构处的防水密封处理。

（5）因建筑层高较低，隔离病房内未设置任何横向风管，空间简洁，医护走廊内未设置任何风管，确保走廊净高合理。

（6）应结合患者的实际需求确定设计方案，而不是凭所谓的经验做设计。许多新冠肺炎患者需要大量吸氧，氧气需求量远大于常规医院中的一般患者，因此氧气源及氧气管道必须进行计算并考虑一定的富余量。

（7）新冠病毒传染性强，建议负压隔离病房通风设备设计考虑备用设施，其他生命支持设备应有一定的冗余。

（8）工程设计周期仅几天，存在少量专业配合及协调不足的问题，在后续现场服务及图纸复查中及时发现和解决问题，设计人员现场指导及协调是此类工程建设的重要环节。

1.7　工程费用分析

1.7.1　费用分析背景说明

为应对新冠肺炎疫情，根据新冠肺炎的临床症状及传播特点，各地政府紧急组织实施的一批应急医院新建及改造工程，为防控疫情传播、保护医疗人员、医治患者、做好保护隔离、对症诊疗提供了基础保障，形成了显著的社会效益。由于疫情防控的紧迫性，项目实施基本无法按照正常建设流程进行，前文以雷神山医院为例就应急临时传染病医院建设问题进行了分析研究，本节从设计特点、施工组织、成本差异等方面分析工程费用的特点，并对有关问题进行探讨，希望对决策审批、资金保障、工程管理、造价咨询等有一定借鉴意义。

1.7.2　项目设计与常规医院费用差异点分析

应急传染病医院为专科类医院，与常规医院相比，在诊疗医护流程上有较大差异。根据病毒种类、传播途径不同，阻断传播采用的方式及标准也有所差异。常规综合类及其他专科类医院用于传染病诊疗易产生感染传播事故，所以，不宜直接用于传染病诊疗。为及时应对疫情、有效控制传播，应急医院的建设和改造就具有较强的现实意义。

1. 建筑设计增量分析

项目选址于原军运会万人食堂及配套停车场，交通便利、设施齐备，场地基本全硬化覆盖，建设条件良好。为防止场地雨水直接排放造成污染扩散，隔离医疗区整体场地加铺抗渗膜及硬化保护层，统一收集并防止雨水下渗。由于项目属于临时应急工程，功能需求明确，以重症患者诊疗为主，所以设计取消了门诊区，住院区病房以医护单元为中轴线呈鱼骨状排列，配套设置药房、医技用房，设置 ICU 病床 61 张及 380m² 的负压检验室。因新冠肺炎临床验证需求，设置有 3 间防辐射 CT 机房。由于新冠肺炎传染性极强，医护人员进出污染区均采取最严格的污染防控流程，卫生通过区域面积在平面布局中占比较大。所有病房设紫外线消毒传递窗以减少接触。病房入口均设置缓冲间，全面设置吊顶以改善负压环境。

2. 结构设计增量分析

为满足临时应急项目的紧迫需求，项目选用轻型模块化钢结构组合房屋作为主体。针对不同功能区不同的开间、进深、层高需求，选用标准轻钢板房作为医护休息单元，选用 3m×6m×2.9m 及 2m×6m×2.9m 的标准集装箱组合为病房单元，选用轻钢结构夹芯板建设层高要求高、承重能力强的医技区域主体。同时由于项目实施时间极为紧张，现场需根据施工单位能组织到的材料进行复核和调整设计。

3. 电气设计增量分析

项目为新冠肺炎诊治专项应急医院，应做好防护，杜绝病毒传播。医院医技区、护理区及病房区须维持全时负压状态，且对供电安全性及稳定性要求特别高。雷神山医院项目供电负荷等级确定为一级，从不同的区域变电站引入 4 路 10kV 高压电源供电，共设置 28 台室外箱式变电站及 11 台室外箱式柴油发电站，总变压器安装容量达到 17720kVA，总发电机安装容量达到 6870kVA，手术室、ICU 等处还设置UPS，医院整体供配电容量达到了普通医院的 2 倍以上。病房、卫生间、走廊、诊室、手术室等需要灭菌消毒的场所设置紫外杀菌灯。考虑项目具有临时性特点，弱电智能化仅设置基本的火灾自动报警及联动控制系统、综合布线系统（包含电话、内网、外网、无线网、设备专网）、病房呼叫系统、安防一键报警系统、视频监控系统、门禁控制系统、病患入口可视对讲系统、ICU 患者监视系统等。

4. 给水排水设计增量分析

考虑项目为传染性很强的呼吸道传染病专项应急医院，给水排水的安全性及重要性需要突出。为防止供水反流污染市政给水网，项目采用断流水箱加压给水泵站供水，泵房出水管设紫外线消毒器，且生活供水设置应急加氯设施，确保供水安全。隔离医疗区为防止交叉感染，室外分设病区污水、非病区污水及室外雨水三套独立排水管网。隔离医疗区污废水经独立管网收集后进入污水处理站统一处理，达标后排放。隔离医疗区室外地面铺设 HDPE 防渗膜，防止带病菌雨水渗入地下污染地下水。因工期要求，项目未设置消火栓及自动喷水灭火系统，按严重危险级配置手提式灭火器，保证 15m 保护范围内均设有手提式灭火器。为防止废水对外传播病毒，病区污水均采用预消毒＋化粪池＋二级处理＋消毒工艺处理，且消毒时间均延长，污水处理站总处理规模为 80m³/h，超过了常规医院污水处理规模。

5. 空调通风设计增量分析

考虑项目为传染性很强的呼吸道传染病专项应急医院，空调系统在控制污染空气排放、保障环境安全方面至关重要。为防止污染空气扩散形成交叉感染，有效保障医护安全，须通过送排风组织形成清洁区、半污染区、污染区的梯度压力差，对病区整体空气流向进行控制。全病区均须设置机械通风系统，且所有区域通风系统的送排风支管上均安装定风量阀，病房送排风支管上同时安装电动密闭风阀。为降低病房空气中的细菌、病毒浓度，呼吸道传染病病房的通风量要求在非呼吸道传染病病房通风量的 2 倍以上，相应送排风机容量均须提高。所有风机均按 1 用 1 备设置，每个送风系统均设置粗效过滤器、中效过滤器、高效过滤器进行三级过滤，排风系统设高效过滤器。根据功能需求，项目设置有氧气、负压吸引、压缩空气等医用气体及管网系统。为满足项目巨大的氧气量需求，共设 6 个储量 20m³ 的液态氧储罐，液氧总储量为 120m³。

1.7.3 项目施工组织费用差异点分析

1. 工人组织管理情况

雷神山医院项目施工工期极短，项目总规模约 8 万平方米，工人需求量极大，而项目的整个建设高峰期正处于防疫管控与春节假期叠加期间，人员组织困难。为保障项目进度，施工总包单位采取了提高人工单价、通过各种渠道募集工人的措施，高峰期现场工人达万余人。由于施工周期极短，无法采用常规的流水节拍方式组织施工，大量交叉作业导致工效降低幅度较大，人工成本巨幅增加。

2. 设备物资供应组织情况

项目实施期间正处于全国防疫管控期，所有工厂基本处于停工停产状态，市场供应均以现有库存为主，且项目实施周期极短，物资设备采购范围辐射全国，货物运输困难，导致项目设备物资价格大幅上升。大宗物资如主体工程所需的集装箱房，雷神山医院、火神山医院两个项目共需近 4000 间，项目所在地区的存量完全无法满足需求，须从北京、上海、广州及

部分西部城市紧急调货，单间运输费即高达数千元。且集装箱房原设计功能以办公为主，现均须改造为病房，并加设传递窗、改造卫生间，改造工程量较大。部分无现货的非标设备须订制生产，工厂非满负荷运转也导致设备成本剧增。

3. 施工机械供应组织情况

受春节假期、疫情防控及交通管制影响，现场机械设备基本为一次性入场，按照施工需要陆续离场，机械降效明显。如为完成大批集装箱房、屋顶风机集中吊装任务，前期即组织大量各类型吊车入场，但前期工作强度较低，待现场集装箱房已基本完成改造工作后，又大批量集中吊装，工作强度极大，施工机械工作降效巨大。

4. 管理及措施成本情况

由于项目工期极短，现场工序存在大量交叉作业，为保证项目顺利正常推进，物资调拨、商务财务、施工管理、质量管理人员数量较常规项目增加数十倍。项目实施期为疫情高峰期，现场人员密集，为进行疫情防控增加了防护费用，如安装测体温系统，发放药品、防护口罩、手套、酒精、清洁用品等防护用品，设置专职防疫医务人员及其办公场区，安排办公区、生活区消杀人员及准备消杀药品等，施工完成后施工人员及管理人员按规定进行隔离，均会产生一定的费用。

1.7.4 项目费用成本探讨

1. 项目成本特点分析

从前面提到的设计及施工组织特点可以看出，应急医院项目相较于常规医院项目增量成本较多，且应急项目相较于常规项目，其实施成本的增加是全方位的。应急项目建设模式与常规项目建设模式完全不同，应急项目的首要目标是进度目标，围绕进度目标开展设计、施工组织，设计工作经常要根据现有资源调整，

经济性并非项目实施的控制性因素。同时，由于项目实施的紧迫性，前期策划工作时间不足，有针对性的成本控制措施不足，导致项目实施工效降低明显，成本增加。

2. 有关建议

从以上成本特点分析，建议从以下几个方面着手加强应急能力建设，提升管控能力，提升工效、降低成本。

（1）加强数字化建设及应用。应用数字化手段可有效提升人员、物资、机械的供应管理能力，优化企业资源，提升应急调度能力，以更经济有效的人员、物资、机械配置完成项目。同时应用数字化技术，可有效保存项目实施过程资料，为后期项目结算、审计等提供有力的基础依据。

（2）提升 BIM 应用能力。应用 BIM 可有效模拟现场施工组织情况，给实际施工组织提供有力参考。提升 BIM 应用能力可更及时有效地为项目服务。特别是实施应急项目模拟，可依据模拟情况优化人力配置，组织流水施工，提升工效。

1.7.5 结语

综上所述，应急医院建设项目特点突出，与常规医院项目相比在费用上存在诸多差异，应急临时医院项目更是差异巨大。因此，在项目前期决策阶段，建议采取常规费用指标加增量费用的计价分析方式。增量费用根据当期的人工、物资市场供应情况综合考虑。在项目实施阶段，建议以成本加酬金方式进行结算，合同中须提前约定好各项成本的确认方式，同时加强审计监督，为项目顺利推进及后期结算打好基础。

2
方舱医院

2.1　武昌方舱医院建设的设计策略

2020 年，当全国人民正在准备迎接春节的到来时，新冠肺炎疫情来势凶猛，感染人数飙升，武汉的医疗资源受到严峻挑战。为阻断疫情传播而建立的方舱医院，使武汉的医疗资源在短时间内得到很大的补充，并与"两山"医院（雷神山医院、火神山医院）、定点医院等医疗机构组成立体防控体系，让疫情迅速得到控制。从某种意义上讲，方舱医院的建设是本次抗疫"战争"的转折点。由洪山体育馆改建的武昌方舱医院（图 2.1、图 2.2）是武汉市首批启用的三座方舱

医院之一，于 2020 年 2 月 5 日顺利启用，该方舱医院从决策建设到首批患者入住仅花费 48h。2020 年 3 月 10 日，武昌方舱医院里的最后一批患者全部出舱。武昌方舱医院是最后休舱的方舱医院，它的休舱标志着武汉市所有的方舱医院全部休舱。

2.1.1　方舱医院概念解析

2020 年 1 月 23 日，武汉市疫情防控指挥部发布 1 号通告，暂时关闭离汉通道，为了应对新冠肺炎疫情，政府制定了应收尽收的策略，在及时治疗患者的同时，隔离感染源，切断传染路径，保护易感染人群。

当时统计数据表明，新冠肺炎患者中，80% 为轻症患者，14% 为重症患者，6% 为危重症患者，根据患者的人群分类，须建设对应的医疗资源，以更好地进行救治工作。根据轻症患者人群数量较大、具备生活自理能力的特点，中国工程院副院长王辰院士提出了方舱医院的概念：在现有公共设施中设置病床和病区，改造城市中大空间的公共建筑，快速建造有遮蔽空间和"三区两通道"的医疗庇护所。王辰院士表示："作为一种战时的简易措施，方舱医院可以在短时间内迅速兴建，大容量集中收治确诊轻症病人。把患者与家人、单位和社区隔离开来，隔离传

图 2.1 洪山体育馆鸟瞰图

来源：百度百科 . 洪山体育馆 [Z/OL].[2022-01-28].
https://baike.baidu.com/pic/%E6%B4%AA%E5%B1%B1%E4
%BD%93%E8%82%B2%E9%A6%86/7974187.

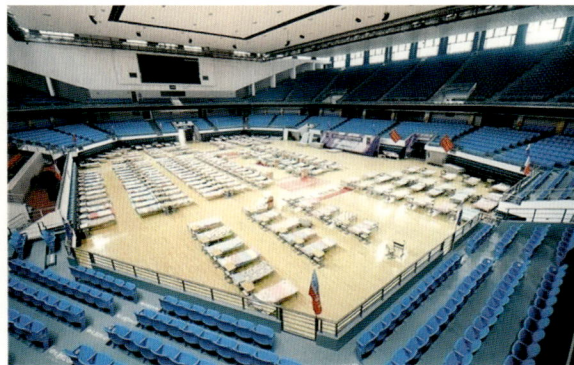

图 2.2 建设中的武昌方舱医院

来源：存档 | 武汉病床 [EB/OL]. （2020-03-15）[2022-01-28].
http://artda.cn/yishubeijing-c-11391.html.

染源，切断传染途径，救治患者，这对于防疫至关重要。"

中南院的技术团队针对方舱医院的需求，将功能需求归纳为分诊、医疗、转诊、生活社交四大功能，并在建筑中予以实现。

2.1.2 建设原则

1. 快速建设

在洪山体育馆中进行改造，建筑本身已具备遮蔽空间和基本的生活设施，只需要进行"三区两通道"的改造，工程量小，可快速建成。一个大型专科医院的建设至少需要 2 年，而武昌方舱医院 2 天即可完成建设。

2. 大规模

洪山体育馆拥有篮球馆、训练馆等大空间，可容纳的床位达 800 张，正好满足收治大量轻症患者的需求。

3. 低成本

武昌方舱医院的改造主要是对"三区两通道"的划分，没有大量的土建工程，相较于同等规模医院的建设费用来说成本极低。

2.1.3 设计策略

2.1.3.1 项目选址

项目选址应该具备以下条件：选址有效覆

盖收治范围，交通便利，市政设施完善，有充裕的设施搭建场地，且与周边重要设施有足够的防护距离。

依照统一安排，武昌方舱医院对口收治武昌区、洪山区、东湖新技术开发区已确诊的新冠肺炎轻症患者。洪山体育馆与这三个区的距离均衡，交通便利，便于患者的快速到达。洪山体育馆满足三个先决条件：大体量、高空间、低层数。大体量可满足大量集中性收治的需求；高空间有较好的空气稀释度，可降低气溶胶传播的风险；低层数则方便患者日常盥洗、物流运送，以及突发情况发生时，集中人流快速疏散。洪山广场是武昌区的中心地段，交通通达性好。洪山体育馆 30m 范围内无高密度住宅、学校，疫情期间邻近建筑中的酒店、餐饮业均已停止营业（图 2.3）。洪山体育馆地势较高，场地平坦，对周边环境的整体影响较小；主场馆与训练馆为

图 2.3 洪山体育馆区位示意图

　　　　　　　　数字化应急医院设计及建造技术

一组建筑，空间较大，并且具备改造"三区两通道"的条件；建筑周边场地开阔，便于方舱车、分诊空间等附属设施的布置；位于武昌区中心，与武汉大学中南医院、武汉大学人民医院等大型综合医院邻近，便于医疗资源的共享。

2.1.3.2 功能的满足

1. 总体规划的布局

（1）功能分区

武昌方舱医院按功能可分为隔离区、医护生活区、分诊治疗区、污物处理区。隔离区位于主场馆和训练馆，处于用地中心；医护生活区位于用地南部；分诊治疗区位于用地西北部；污物处理区位于用地西北部。武昌方舱医院具体功能分区如图2.4所示。

（2）出入口的设置

洪山体育馆交通便利，原用地内有较多的出入口，大部分都可以供车辆出入。改造时按功能需求开设出入口：南面开设医护人员出入口，用于医护人员和清洁物资的进出；北面为患者入口；西北角设置污物出口。

（3）流线的组织

流线设置注意洁污流线严格分开。用地北面为污染区，设置患者流线和污物流线，用地

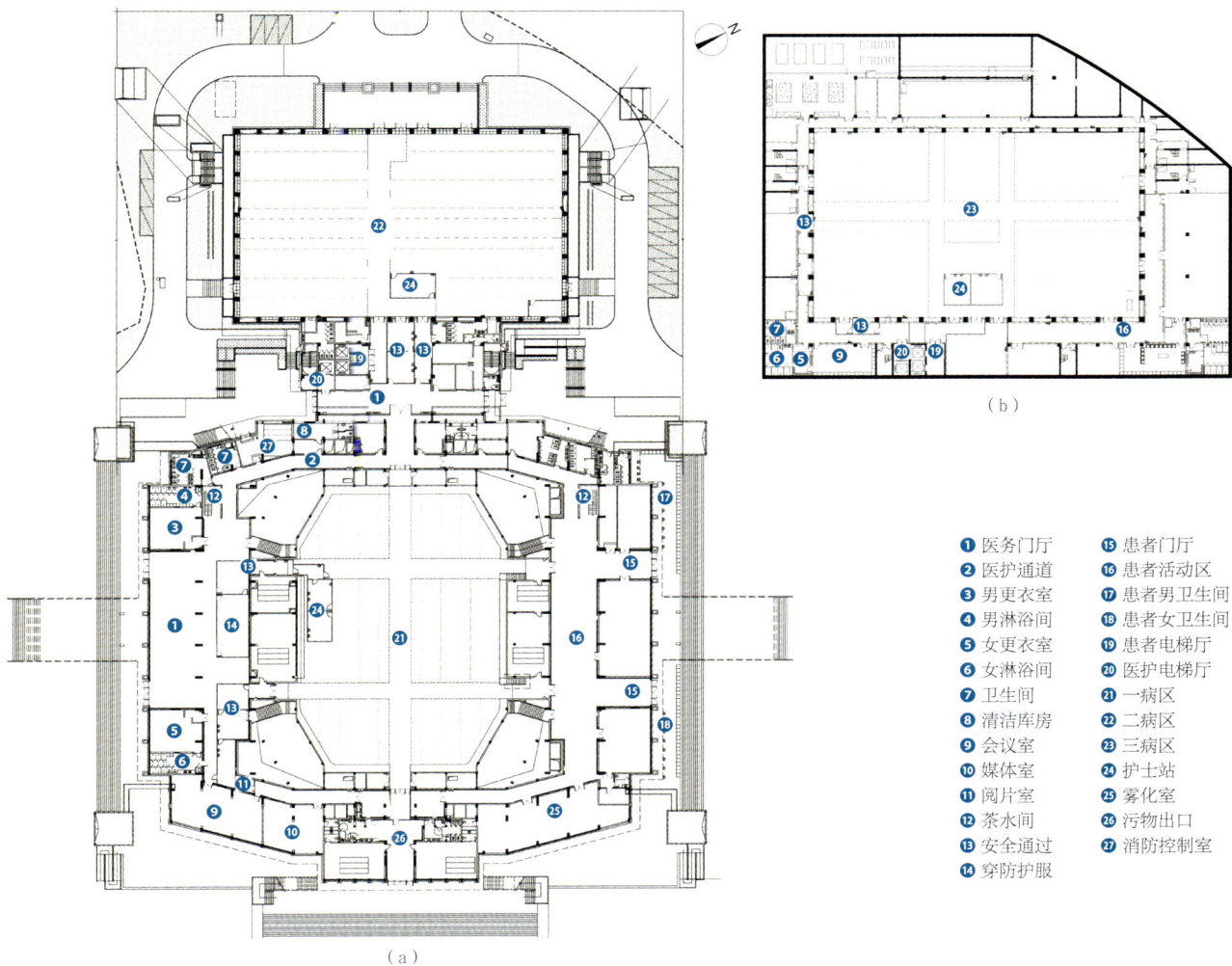

（a）

① 医务门厅　⑮ 患者门厅
② 医护通道　⑯ 患者活动区
③ 男更衣室　⑰ 患者男卫生间
④ 男淋浴间　⑱ 患者女卫生间
⑤ 女更衣室　⑲ 患者电梯厅
⑥ 女淋浴间　⑳ 医护电梯厅
⑦ 卫生间　　㉑ 一病区
⑧ 清洁库房　㉒ 二病区
⑨ 会议室　　㉓ 三病区
⑩ 媒体室　　㉔ 护士站
⑪ 阅片室　　㉕ 雾化室
⑫ 茶水间　　㉖ 污物出口
⑬ 安全通过　㉗ 消防控制室
⑭ 穿防护服

图2.4　武昌方舱医院平面图
（a）一层平面图；（b）地下一层平面图

南面为清洁区，设置医护流线和供应流线。分诊治疗区位于用地西北面，为室外的清洁区与污染区的交界处。

2.建筑功能的设置

（1）"三区两通道"的改造

"三区两通道"的改造是武昌方舱医院改造的核心。洪山体育馆原本为体育建筑，通过现场的详细考察，基本具备改造为方舱医院的条件。

①分区。

医护区。医护区是清洁区，从用地南面的医护人员出入口进入建筑。三个隔离分区的医护区通过走道和垂直交通连成一个整体。医护区内设置医护办公室、休息室、值班室、物资库房，并利用原有的运动员沐浴更衣室改造出医护人员的消洗更衣区域。三个隔离分区均有自己的医护区，但是医护区又相互联系，实现资源共享，大大提高了使用效率。

患者区。患者区为污染区，从用地北面进入建筑内部，分三个区域，每个区域都有自己独立的出入口。因患者区的污废水须单独处理，所以将患者区的卫生间封闭，采用集成的卫生间、淋浴房、洗手盆，在室外设置临时用水区域，确保武昌方舱医院用水的生物安全。患者区内用物理隔断将病房划分开，形成 4 个病床单元，便于建设和管理。

安全通过。安全通过是半污染区，是保证武昌方舱医院生物安全的核心，这一部分是原洪山体育馆没有的，根据需要加建。针对医护进出患者区必须设置单一流线，进出口分开，并保持一定距离，还需要设置穿脱隔离服、防护服的空间及缓冲间。因武昌方舱医院每个分区的床位数较多，设计时加大了安全通过空间的面积。水、电、风是安全通过发挥作用的重要保障。电的需求问题基本可以解决，水和风的需求问题解决难度较大。在用水方面，须充分利用现有资源，通过靠近原有卫生间设置需要用水的房间。在通风方面，需要后期加建通风设施。需要说明的是，当时武昌方舱医院因建设时间短，而且在春节期间建设，设备采购存在问题，专业的施工队也很难寻找，在几乎准备放弃的情况下，还是坚持把通风设施做出来了，从后期的运行情况来看，通风设施必不可少。

建成后，武昌方舱医院实景如图 2.5 所示。

②流线设计（图 2.6）。

医护流线。医护人员从用地南面的清洁区域进入医院，在更衣区域穿防护服、隔离服后，经过缓冲间，进入患者区。完成工作交接班后，经另一出口设置的脱隔离服、防护服的区域离开患者区域。出口的附近设置有沐浴更衣区域，可进行彻底的清洗消毒。

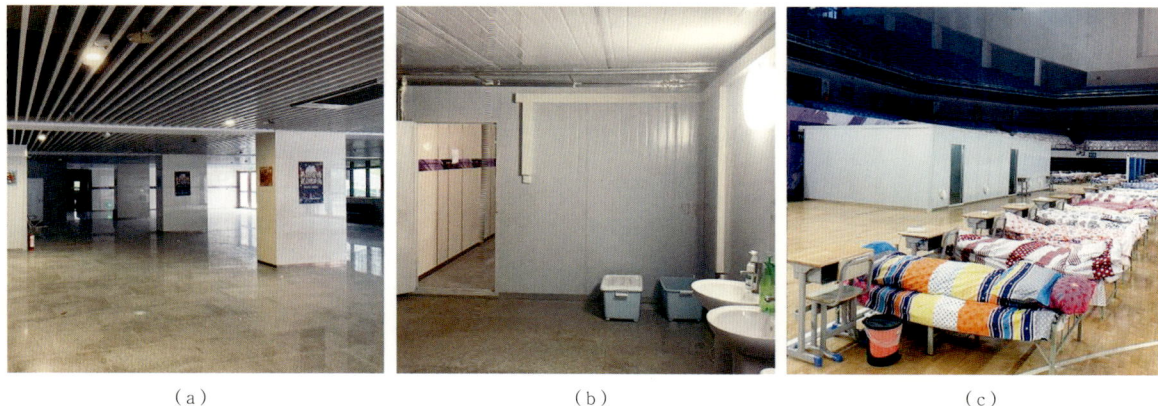

| （a） | （b） | （c） |

图 2.5　武昌方舱医院实景
（a）清洁区；（b）半污染区；（c）污染区

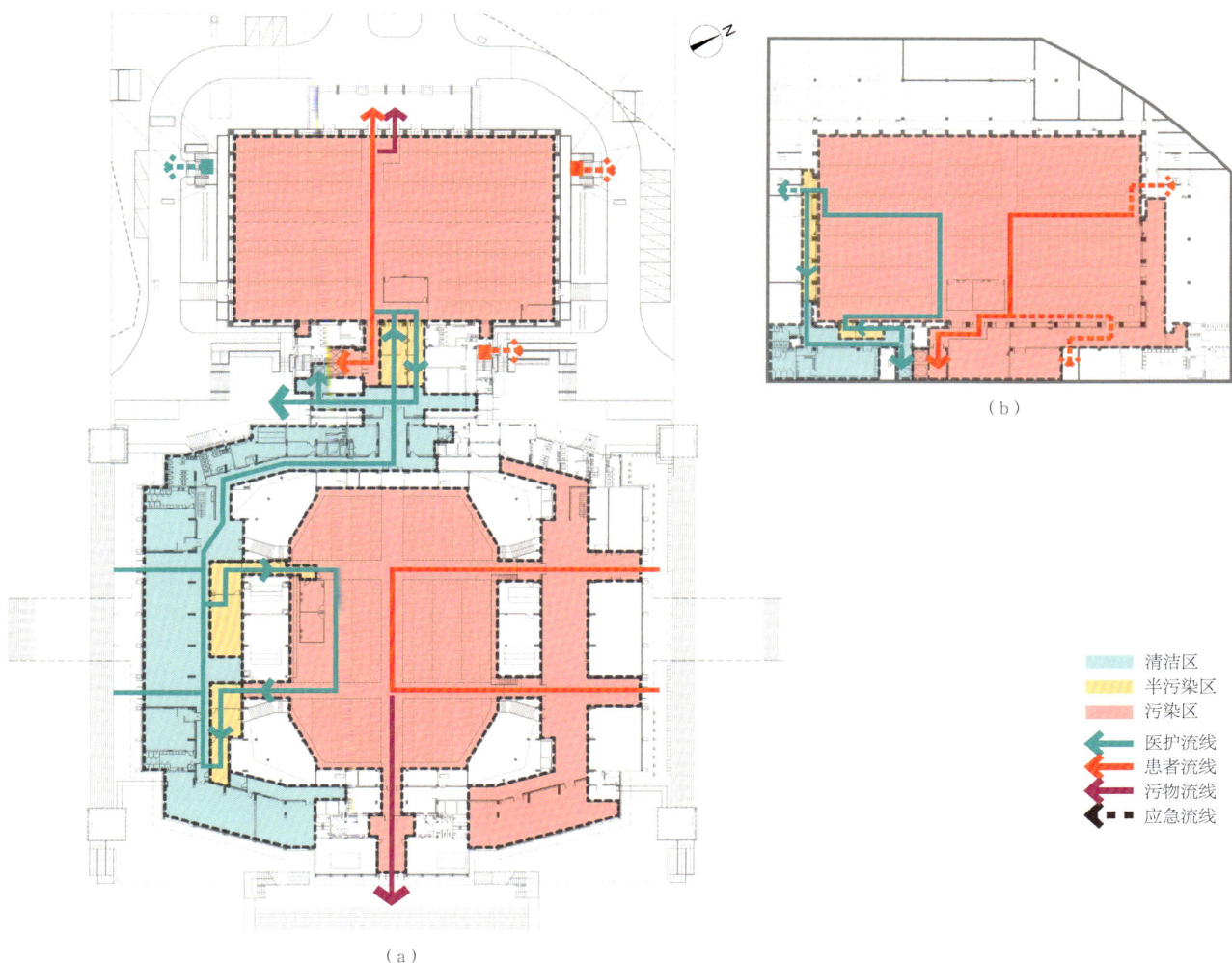

（a）

（b）

清洁区
半污染区
污染区
医护流线
患者流线
污物流线
应急流线

图 2.6　武昌方舱医院流线分析图
（a）一层平面图；（b）地下一层平面图

患者流线。患者统一由用地北面的入口进入医院，先到用地西北面设置的分诊治疗区进行分诊，分诊完成后，再进入对应的患者区。医院内的生活物资由专人配送至床头，患者解决卫生问题时由原入口利用地西北面的集中用水区域。患者康复后，经过严格的检查消毒后，从东面设置的出口离开院区。

污物流线。每个患者区均设置独立污物出口，在污物出口附近设置污物暂存间，医疗垃圾和生活垃圾在污物暂存间集中收集后，统一由专人运输至焚烧炉处理。

应急流线。应急流线能保证改造后的建筑满足使用人员安全疏散的需求，而且能保证人员疏散的有序性。

（2）应对实际问题

①设计分区封堵。清洁区、污染区的划分是实现安全隔离最重要的手段之一。设计采用岩棉彩钢板、聚氨酯泡沫胶等材料进行全方位分隔。需要注意的是，增加的隔墙顶部，往往设有吊顶造型及各类管线，给封堵施工造成了不少困难（图 2.7）。因此在划分区域时，应尽量结合原建筑防火分区及隔墙设置，保证清洁区、污染区的绝对分隔，确保安全（图 2.8）。

②适用的安全通过。在清洁区与污染区之

间，设立两侧均有门的小房间，作为医护人员进出污染区的安全通过（图2.9）。医护人员由清洁区经一次更衣、二次更衣进入污染区，离开时先通过缓冲间，经一次脱衣、二次脱衣后，返回清洁区。安全通过区域应设置用水设施（图2.10）。应结合病区规模、场地尺寸及医护人员人数等条件，合理设置安全通过的个数、位置、面积，以避免出现距离过远、进出排队时间过长等问题。

③合理安排病区床位。病区内的病床应分组设置，每组床位数应小于50张，组外主通道宽度尽量大于3.6m（图2.11）。病区内的视线应开阔，便于医护人员的工作。每组之间设置隔断，提高患者的私密性（图2.12）。可采用高低床，既方便床头插座布线，也为每床提供了一定的储物空间（图2.13）。

病区护士站均布置在病区中部，同时靠近安全通过房间；如果病区面积过大，应考虑分设护士站，以提高护理效率，减少无用的往返。

2.1.3.3 快速建造

快速建造是应急医疗设施建设必须达成的目标，大量的建设人员云集现场，必须有一套科学完善的方法，才能保证建造的快速性。

1. 功能选择

洪山体育馆于1986年对外开放，2006年为满足全国第六届城市运动会的需要进行改造升级，是湖北省第一座大型、多功能的体育馆。本次改造选择洪山体育馆空间最大的主体育馆和附属的训练馆。洪山体育馆的改造围绕"三区两通道"展开，即清洁区、半污染区、污染区，以及医务人员通道、患者通道。患者居住的场馆为污染区；场馆外围的环形办公区域为清洁区，医护人员可以在此休息，医疗物资也在此保管；而半污染区则为穿衣间和脱衣间。

利用原有建筑改造隔离病房区、后勤保障区，需增加的功能单元有：安全通过单元、医护治疗单元、污物处理单元、患者用水单元、分诊单元。

图2.7 顶部开敞时有气溶胶传播风险

图2.8 清洁区与污染区由多道墙体分隔

图2.9 宽敞的安全通过

图2.10 所有安全通过房间均设置洗手池

图2.11 患者公共区

图2.12 分区隔墙

图 2.13　高低床设置　　　图 2.14　室外患者卫生间　　　图 2.15　方舱车及室外帐篷

2. 建造材料的选型

室内改造的区域包括安全通过单元、医护治疗单元、污物处理单元内的污物暂存间，为增设空间。选用岩棉彩钢板作为隔墙材料，这种材料有足够的强度　并且具备隔声、防火、便于清洁的特性，在建造上也便于拼装，便于获取，技术成熟。患者用水单元在室外重新设置，使用集成的卫生间（图 2.14）、化粪池和成品的洗手盆，现场拼装、摆放。分诊单元中的分诊室采用帐篷搭建（图 2.15），在方舱车旁选择水电条件便利的硬质场地搭建。

图 2.16　留言墙

2.1.3.4　人文关怀

由于武昌方舱医院容纳的患者较多，比专业医疗建筑简陋，而轻症患者大部分有行动自理能力，确诊、转院、长期处于封闭环境，都容易产生焦虑心态。在建筑设计上尽量考虑周全，体现人文关怀。

在保障病床数量的情况下，尽量加大每个病区的公共面积，利用原建筑的门厅、公共通道，为患者间的生活社交提供场所；尽量改善供暖、通风条件，为每床提供充电设备，尽量提升室内环境和生活品质；制作建筑使用说明书，张贴平面流线图和房间挂牌。在建筑使用过程中，公共区域设置了留言墙，张贴了国旗、党旗，张贴了抗疫艺术画作，预留足够的活动区域，让大家可以开展一些社交活动，如跳广场舞等（图 2.16 ～图 2.18）。

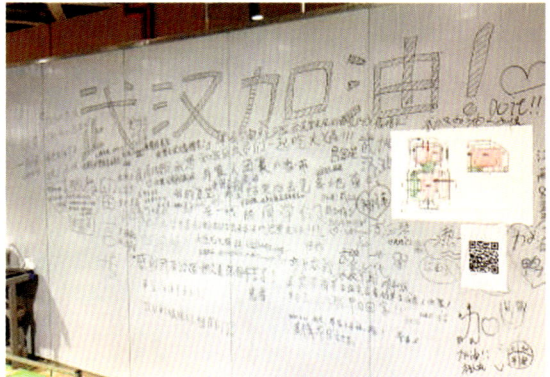

图 2.17　张贴的国旗

图 2.18　在病区内跳广场舞
来源：武汉方舱医院广场舞 [Z/OL].（2020-02-15）[2022-01-28].
https://www.thepaper.cn/newsDetail_forward_6027756.

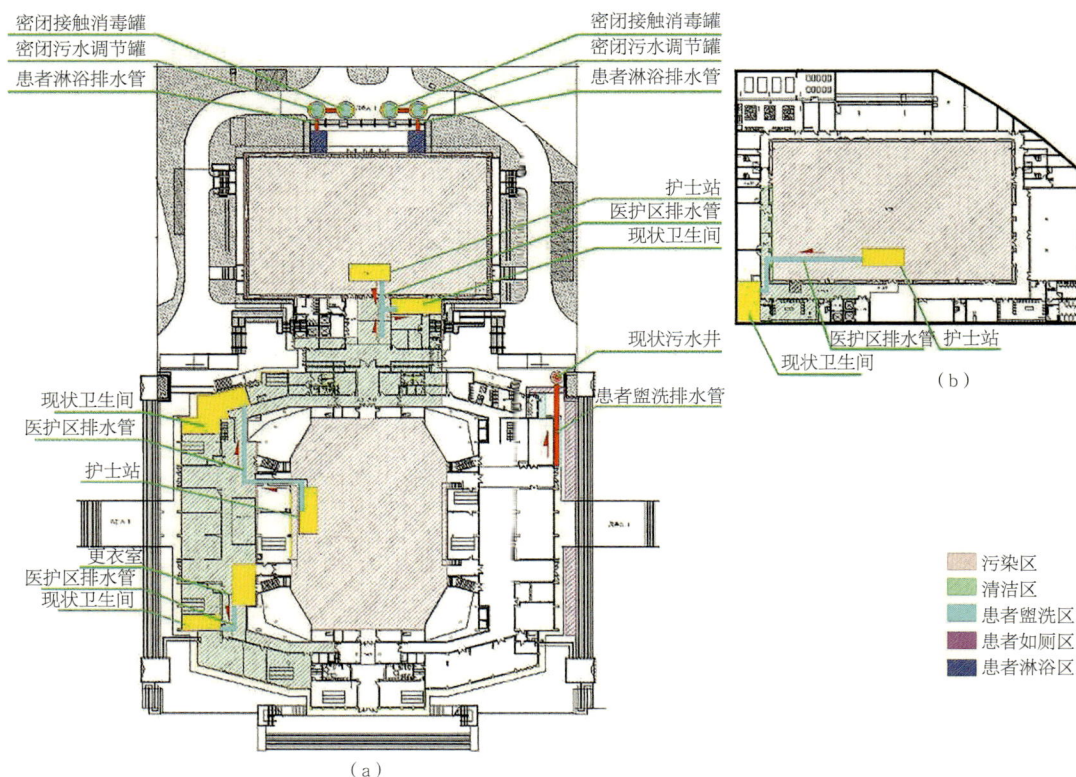

图 2.19 医护区、患者区排水示意图
（a）一层平面图；（b）地下一层平面图

图例：
- 污染区
- 清洁区
- 患者盥洗区
- 患者如厕区
- 患者淋浴区

2.1.3.5 设备保障

1. 给水排水设计

根据建筑的功能分区，医护人员使用场馆内固定的厕所、淋浴间，并在护士站、更衣室增设洗手盆供医护人员使用；患者则使用场馆外的临时移动厕所、盥洗池、淋浴间。场馆内污水由场馆内现有污水管道系统收集，场馆外污水由新建室外架空管道系统收集，两套污水收集系统完全独立。医护区、患者区排水示意如图 2.19 所示。

（1）场馆内污水收集系统。在建筑完成面上直接敷设排水管，收集护士站、更衣室洗手盆污水，就近排至场馆卫生间排水系统，场馆内既有排水系统不做改动。

（2）场馆外污水收集系统。患者盥洗区、如厕区、淋浴区污水采取不同的方式收集。

①盥洗区洗手盆布置在篮球馆入口平台，污水就近接入现状污水检查井。

②临时移动厕所设置在篮球馆入口平台，污水进行预消毒处理后由吸粪车外运至异地处理。

③临时淋浴间设置在羽毛球馆入口平台，污水利用场地高差自流至平台下密闭接触消毒罐。淋浴间污水外运至异地处理。

（3）污水处理系统。场馆内污水、患者盥洗污水排入现状室外污水管网，进行一次消毒，在市政管网入口处污水检查井进行二次消毒处理后排放，并设置在线监测仪对余氯及 pH 值进行监测。

2. 通风设计

为了保证医护人员、工作人员和患者的健康安全，尽量降低武昌方舱医院对周边环境的影响，空调通风系统的改造原则如下。

（1）清洁区和污染区的空调系统独立运行。

（2）清洁区、半污染区和污染区保持一定的压力梯度且压力逐步降低。

（3）污染区及半污染区的排风经过处理后方可排放（图2.2、图2.21）。

空调通风系统改造尽量利用原系统设施设备，以快速、易行的方式进行改造。

（1）改造三个病区的单元系统，保证病区单元保持负压状态。

（2）在清洁区和病房区域设置若干台具有杀菌消毒功能的空气消毒机，净化室内空气（图2.22）。

图2.20　主场馆二层室外平台安装粗效、中效、亚高效过滤器

图2.21　训练馆屋顶安装排风机及粗效、中效、亚高效过滤器

图2.22　病房区安装的空气消毒机

3.电气设计

洪山体育馆电气改造以配电安全为核心，兼顾合理性及易于改造性，在满足武昌方舱医院用电需求的同时，设计亦考虑方便设备运维。

（1）武昌方舱医院共有3个病区，分别位于主场馆一层、训练馆一层及训练馆地下一层。共新增6个区域配电箱。

（2）隔离区病床床头设置插座供患者使用（仅供手机、台灯、电热毯等小功率用电设备使用），按每床1个插座、1个插线板设计。

（3）3个病区的照明采用原场馆设计的智能照明系统。

（4）主场馆东边出口处设置小厨宝电源，就近自配电箱备用回路接线。

（5）场馆原设计Wi-Fi全覆盖，可为医患提供互联网连接。

（6）安全注意事项如下：

①床头插座可供电热毯使用，应提醒患者在不必要开启时注意关闭电热毯开关，以减少火灾隐患；

②插座、插线板等设备应避免水泼溅，减少配电回路的短路隐患；

③由于改造时间有限，训练馆地下一层夹层主供电电缆均沿地面明敷，应避免非专业运维人员进入该区域，以避免发生触电事故。

2.1.4　结语

2020年3月10日下午5点，随着最后一批49名患者从洪山体育馆走出，武汉市首批启用的方舱医院之一——武昌方舱医院正式休舱（图2.23）。方舱医院，这一在武汉市新冠肺炎疫情防控关键时期发挥了关键作用的特殊医院，圆满完成了历史使命。

方舱医院在本次抗疫中发挥了重要作用，对将来的城市建设有着重要的意义。武昌方舱医院作为最先启用的三座方舱医院之一，在建

图 2.23　武昌方舱医院休舱仪式现场

来源：洪山体育中心　提供

图 2.24　电热水器安装实景图

设时没有可借鉴的经验，面临的压力和困难也最多。回顾武昌方舱医院建设和使用过程，总结抗疫经验，编制设计导则，形成统一的标准，可为后续的方舱设计提供较完整清晰的指导。

将来在新建的大型公共建筑项目中，应加强防灾避险的专项设计，统一管理，充分发挥公共建筑的价值，实现快速的平战转换，加强现有医疗建筑的兼容性，预留地下室、室外空间改造的可能性，做好安全储备，从而应对将来可能出现的各种未知风险。

2.2　方舱医院给水排水及消防系统设计——以江夏大花山方舱医院为例[9]

本节内容以江夏大花山户外运动中心乒羽馆改造工程为例讲解方舱医院的给水排水及消防系统设计。

2.2.1　工程概况

江夏大花山户外运动中心乒羽馆于 2019 年 8 月竣工并投入使用，有完善的室内外给水排水、消防系统。为解决新冠肺炎确诊轻症患者收治问题，将乒羽馆西区一层乒乓球馆、二层羽毛球馆改造成方舱医院，总床位约 628 张。

2.2.2　给水系统

江夏大花山方舱医院最高生活日用水量为 198.4m³。根据《新型冠状病毒肺炎应急救治设施设计导则（试行）》[10] 的要求，给水系统宜采用断流水箱供水，江夏大花山方舱医院若改造时采用断流水箱 + 供水泵的给水系统，考虑疫情时期一体化给水泵站运输、安装等因素，不利于工程的快速实施。

本改造工程用于收治轻症患者，回流污染风险较低，给水管道从市政给水管道引入，采用减压型倒流防止器防止回流污染。所有洗手盆的水龙头均采用感应水龙头，蹲便器采用脚踏阀，大便器选用冲洗效果好、污物不易黏附在便槽内且回流少的器具。所有卫生器具的用水效率等级为二级。

2.2.3　热水及饮用水系统

1. 热水系统设计

江夏大花山方舱医院医护区淋浴间设有 3 个淋浴器，考虑工程建造时间，采用容积式电热水器供应热水（图 2.24），每个淋浴器设置 1 台储热容积为 80L 的电热水器。电热水器水温稳定且温度可调节，确保舒适好用；采用水电分离技术，确保安全。

结合其他方舱医院建设情况，后期运行存

在增加淋浴器数量的情况，设计中有条件的可设置空气源热泵集中热水系统，此外可设置电热水炉集中热水系统，一般1台90kW的电热水炉可以满足9个或10个淋浴器的用水需求。

2. 饮用水系统设计

每个护理单元应单独设置饮用水供水点。江夏大花山方舱医院设置带过滤功能的电开水器，保证医患人员喝到安全健康的饮用水。对于电开水器数量的确定，除考虑满足患者正常喝水的需求外，还需要考虑患者会使用此部分开水兑冷水后擦拭身体，对开水的需求量较大，应选择大功率的开水炉（12kW以上）。

电开水器建议靠近卫生间，开水器的排水可就近接入卫生间地漏。

2.2.4 排水系统

1. 污水的分类

江夏大花山方舱医院最高日排水量为198.4m³。针对医护人员和患者分设用水点，医护人员使用现有的固定厕所和医护入口处工作人员专用移动厕所，患者经专用通道使用隔离区的移动厕所。方舱污水主要分为医护人员生活污废水和隔离区污废水 两种污水严格分流排放。

2. 污水的收集和消毒

医护人员生活污废水排至已有的室外污水检查井，在化粪池投加消毒剂进行一级消毒。隔离区污废水经管道利用重力自流至污水储罐，在污水储罐进水口处进行一级消毒，污水在污水储罐内停留至少1.5h，经处理后的污废水排至化粪池，和医护人员生活污废水汇合后进行二次消毒。

江夏大花山方舱医院对移动厕所的排水方式进行了论证，常用的排水方式有两种：采用吸粪车抽排[11]和设管道收集至污水储罐。隔离区设置35个移动厕所，采用吸粪车抽排工作量大，并且须人工投加消毒药片。在疫情时期，应采取适合改造、花费人力物力较少的方式。设计中架高移动厕所，污废水依靠重力自流至室外埋地储罐，污废水管道明敷在地面上，做好管道保护措施，减少管道开挖工程量，对现状破坏较小，便于疫情结束后恢复现场（图2.25）。

污水储罐有效容积应满足污废水停留时间不少于1.5h的要求。江夏大花山方舱医院项目设计病床为628张，最高日用水量取300L，时变化系数取2.5，污水储罐设计最小有效容积为29.5m³。污水储罐一般有推流式和间歇式两种（图2.26），考虑到工程实际中污水储罐需要快

图2.25 移动厕所污水预消毒处理及外运实景
来源：王涛、吴平、秦晓梅，等.方舱医院污水收集处理系统现状及对策分析[J].给水排水，2020.56（5）：22-26.

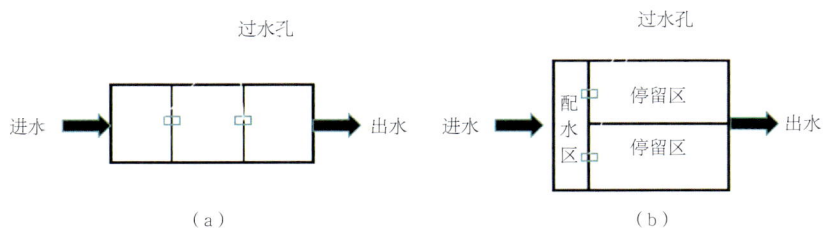

图2.26 污水储罐构造形式
（a）推流式储罐；（b）间歇式储罐
来源：秦晓梅，胡颖慧，危忠，等.方舱医院给水排水及消防系统设计——以江夏大花山户外运动中心乒羽馆改造工程为例[J].给水排水，2020，56（4）：32-35.

速安装使用，设计选用成品化粪池作为污水储罐（图 2.27）。成品化粪池单个有效容积 40m³，设 2 个。为保证污废水有足够的停留时间，可容纳的污泥量应小于 50m³，每天产生的污泥量约为 0.211m³，最大清掏周期可达 230 天，满足改造期间的使用要求。移动厕所（图 2.28）的生活污水与洗浴区生活排水必须经过消毒处理，达到生态环境部颁发的《新型冠状病毒污染的医疗污水应急处理技术方案（试行）》[12] 的相关要求后排放。武汉市应按 2020 年 2 月 4 日市生态环境局、市卫生健康委、市水务局、市城管执法委印发的《关于做好全市方舱医院医疗污水处理有关工作的紧急通知》[13] 的要求进行消毒。采用液氯、二氧化氯、次氯酸钠、漂白粉或漂白精等消毒剂实施消毒时，污废水在消毒接触池的接触时间不应低于 1.5h，余氯量大于 6.5mg/L（以游离氯计），粪大肠菌群数少于 100 个 /L，参考有效氯投加量为 50mg/L。

对于消毒剂的选择，液氯虽然消毒效果好，但属于剧毒危险品，存储氯气的钢瓶属高压容器，有潜在危险。考虑到药剂消毒能力和使用方便程度，在污水储罐和化粪池投加浓度为 10% 的次氯酸钠溶液，采用自动投药方式（图 2.29），次氯酸钠溶液用成品次氯酸钠现场制备。

3. 通气管的设置和消毒

江夏大花山方舱医院项目设计中有两类通气管，第一类为管道通气管，第二类为设备通气管（主要用于移动厕所、污水储罐和化粪池

图 2.27　污水储罐实景

来源：秦晓梅，胡颖慧，危忠，等 . 方舱医院给水排水及消防系统设计——以江夏大花山户外运动中心乒羽馆改造工程为例 [J]. 给水排水，2020，56（4）：32-35.

图 2.28　移动厕所实景

来源：秦晓梅，胡颖慧，危忠，等 . 方舱医院给水排水及消防系统设计——以江夏大花山户外运动中心乒羽馆改造工程为例 [J]. 给水排水，2020，56（4）：32-35.

图 2.29 污水消毒投药设备实景

来源：秦晓梅，胡颖慧，龙忠，等 . 方舱医院给水排水及消防系统设计——以江夏大花山户外运动中心乒羽馆改造工程为例 [J]. 给水排水，2020，56（4）：32-35.

通气）。管道通气管利用现有的卫生间通气管，通气管位置已经固定，应与暖通专业配合，保证新风取风口远离管道通气管出口。设备通气管为设备自带，应设置在通风良好且不影响通行的地方。

管道通气管和设备通气管管口均应设置高效过滤器或其他可靠的消毒设备。市场上的高效过滤器初阻力约为 250Pa，通气管需要进气和排气，难以克服高效过滤器的阻力，高效过滤器只能起隔离的作用，并不能杀死病毒，因此采用紫外线消毒设备对通气管口进行消毒。紫外线消毒设备具有消毒能力强、设备便于采购安装等特点，非常适合本工程。

2.2.5 消防系统

根据应急管理部消防救援局下发的《发热病患集中收治临时医院防火技术要求》[14]规定：任一层建筑面积大于 1500m² 或总建筑面积大于 3000m² 的临时医院应设置自动喷水灭火系统。临时医院应按严重危险级场所配置灭火器。

根据湖北省住房和城乡建设厅印发的《方舱医院设计和改建的有关技术要求》[15]规定：原有消防设施设备能正常使用；应按严重危险级场所配置相应数量灭火器；贵重设备用房、病案

室和信息、中心（网络）机房应设置气体灭火装置；方舱医院内若增设生活给水系统，且原建筑室内消防系统未配置消防软管卷盘时，可增设消防软管卷盘或轻便消防水龙头，其布置应满足同一平面至少有 1 股水柱能达到任何部位的要求；护士站宜配置微型消防站，移动式高压细水雾灭火装置贮水量宜为 100L。

江夏大花山方舱医院在原有体育场馆基础上改造，原有建筑已设置消火栓、自动喷水灭火系统、灭火器等消防设施，改造时利用原有的消防设施，并保证消防设施不被遮挡。因移动式高压细水雾灭火装置采购周期长，江夏大花山方舱医院工程未设置。为了保证消防安全，江夏大花山方舱医院工程按照严重危险级场所加密布置灭火器，针对体育馆等大空间设置推车式灭火器 MFT/ABC（20kg，6A），保护距离按照 20m 控制，其他医护办公区域设置手提式灭火器 MF/ABC（5kg，3A），保护距离按照 10m 控制。

有的体育馆因为建造年代久远，未设置自动喷水灭火系统，在改造成为方舱医院时无法满足应急管理部消防救援局的相关规定，此时应与消防主管部门沟通，利用原有的消防设施，采用其他加强措施，如加密灭火器的布置等。

2.2.6 管材选用及敷设

室内外生活给水管道均采用 PPR（无规共聚聚丙烯）给水管道，热熔连接。为施工便捷，室外给水管道明敷在绿化带中，在医护区就近接至已有的生活给水管道。室内给水管道明敷于墙壁上，管道不变径，减少管道规格，便于采购与施工。

室内外增设的移动厕所，均采用架空的方式布置，管道明敷在地面，尽量减少管道开挖。室外污水排水系统采用室内管道连接的方式，并应当设置通气管和清扫口。排水管材选择 PE 管，热熔连接。若因条件所限，采用承插式接口的排水管，要考虑防止接口漏水的措施。

2.2.7 总结

1. 生活给水系统

对于生活给水系统选择，建议有条件的情况选择断流水箱供水，优先选用室外一体化给水泵站，便于施工安装，同时水箱预留应急补水口和应急加氯口，保证紧急情况下的水质和水量。

2. 生活热水系统

对于生活热水系统，建议有条件的设置空气源热泵集中热水系统，其次考虑设置电热水炉集中热水系统。

3. 生活排水系统

医护区污水和隔离区污水的排水应严格分开，分别按二级和一级消毒处理，并充分利用已有化粪池作为消毒接触池。

4. 消毒

对于管道及设备通气管的消毒，建议采用紫外线杀毒器，不建议采用高效过滤器。对于污废水的消毒，消毒剂建议采用成品二氧化氯溶液或成品次氯酸钠溶液，不建议采用液氯以及需要现场制备的二氧化氯溶液和次氯酸钠溶液。

5. 消防系统

对于大空间场所，建议设置推车式灭火器。

6. 管材选用及敷设

排水管建议采用塑料管材，宜采用热熔、电热熔连接，不宜采用承插连接。给水排水管道的敷设建议采用明敷，做好防护措施。

2.3 电气设计要点

本节内容以中国光谷日海方舱医院改造建设为例讲解方舱医院的电气设计要点。

2.3.1 项目背景与项目简介

为应对新冠病毒疫情下床位不足的情况，应东湖高新区管委会要求，将湖北日海通讯技术有限公司厂区改造为收治新冠肺炎轻症患者的临时方舱医院，即中国光谷日海方舱医院。

项目位于武汉市江夏区藏龙岛，原有功能为工业厂房，厂区总占地面积 188864m²，总建筑面积 287500m²。将其中 4 栋厂房（分别为钣金车间、机房事业部、注塑车间、杆塔制造部）改造为方舱医院，满足临时收治新冠肺炎轻症患者的要求。其中，B 号方舱（原钣金车间）改造面积约 9984m²，共计 610 床位；C 号方舱（原机房事业部）改造面积 16156m²，共计 1445 床位；D 号方舱（原注塑车间），改造面积 8319m²，共计 654 床位；E 号方舱（原杆塔制造部）改造面积 12081m²，共计 980 床位。以上方舱均为单层建筑（图 2.30）。配套改造围墙及其范围内的配套设施，包含垃圾暂存间、物资库、移动 CT 设备等。

中国光谷日海方舱医院的总平面遵循医患分离的设计原则[16]，医护及管理出入口位于厂区原主要出入口，病患及污物出入口位于基地西南角。

根据使用需求及厂房大空间特点，每个方舱分为 3 个区域：①患者洗漱区，位于厂房外部。②医护区，位于厂房入口。医护区内部，在清洁区设置了男女值班室、耗材库、药品库、仪器库、会诊室、强弱电井、监控室；在半污染区设置医生办公室、医疗器械处置室、护士站。③患者区域，分为留观治疗区（主要为隔离床位）、活动区。

中国光谷日海方舱医院设计及改造进度计划如下：

2020年2月4日，武汉市指挥部决定再建8座"方舱医院"；

2020年2月5日，中南院接到中国光谷日海方舱医院设计任务；

2020年2月5日晚，中南院提供初步设备材料表供施工单位备料；

2020年2月6日晚，中南院完成施工图设计；

2020年2月20日，中国光谷日海方舱医院C号方舱通过验收，交付使用，后续各舱也陆续交付。

经现场踏勘，各厂房现状与图2.31所示基本相同。

（1）供电：厂房原有2路10kV市政电源进线，其中厂房2号配电房设有2台变压器为改造的4栋厂房供电，变压器容量分别为1250kVA及1000kVA，2台变压器设置有联络开关。室外另有一台容量为630kVA的箱式变压器为注塑车间部分工艺负荷供电。厂房内部各区域分布有电源配电箱，配电箱进线电缆及开关基本完整，可作为改造的电源总箱。

（2）照明：车间照明灯具部分可用，照明线路部分缺失，照明配电箱分散设置于厂房内钢柱上，照明灯具及配电系统需重新设计。

（3）防雷设施：原防雷接地装置设计完备，可利用原有防雷装置，不做调整，仅在新增带淋浴功能房间增设局部等电位联结。

（4）火灾报警系统：厂房内消火栓报警按钮完好，无其他火灾报警设施，本设计不做改动。

（5）其他弱电系统：无其他弱电系统，须根据需要新增。

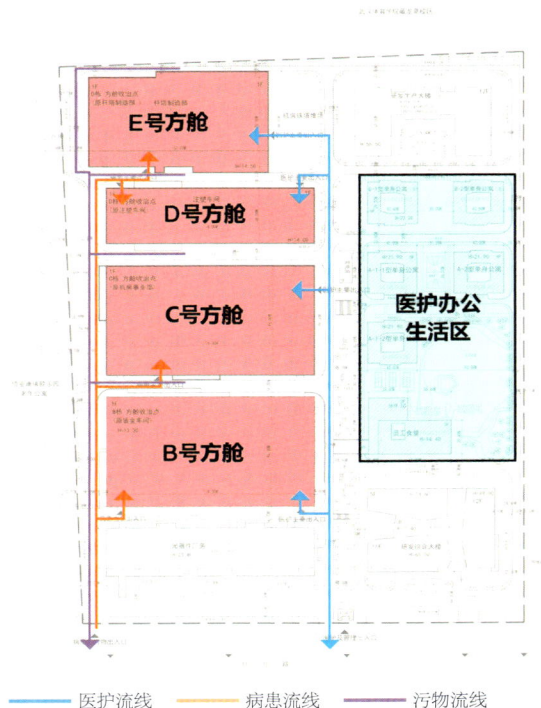

医护流线　　病患流线　　污物流线

图2.30　中国光谷日海方舱医院改造总平面图

图2.31　C号方舱改造前内部

2.3.2　供配电系统

中国光谷日海方舱医院属于临时医疗建筑，需要遵守医疗建筑相关规范，但是它的建设周期很短，需要迅速建成并投入使用，因此，要在不违反规范的情况下，综合考虑施工工期及设备采购的限制，电气系统应简单可靠，设备选型应便于采购安装。

根据《医疗建筑电气设计规范》（JGJ 312—2013）及《传染病医院建筑设计规范》（GB 50849—2014），结合使用需求，确定涉及改造的供电设备负荷分级见表2.1。

涉及改造的供电设备负荷分级　　　　　　　表 2.1

负荷等级	用电设备名称
一级负荷中特别重要负荷	应急照明及疏散指示系统；监控中心、智能化系统机房；医护区正压送风机
一级	治疗室、护士站等场所的诊疗及照明设备
二级	诊断用电设备（CT 等）
三级	普通空调等一级、二级负荷之外的其他负荷

注：表中负荷等级根据中国工程建设标准进行划分。

　　根据各专业提资，用电设备容量为 B 号方舱 1400kVA、C 号方舱 1700kVA、D 号方舱 1400kVA、E 号方舱 1500kVA，主要为隔离病区插座用电，总计算容量为 2200kVA；原有变压器容量可满足用电需求。

　　改造中增设 1 台 800kW 室外柴油发电机（图 2.32）为一级负荷中特别重要负荷及一级负荷供电。柴油发电机自带油箱且预留接油口。为了减少对原有供电系统的大幅调改，节约施工工期，采用柴油发电机并入市电母线的方式供电，接入市电母线的柴油发电机出线开关与变压器出线及联络开关连锁，仅当变压器出线开关均断开时方可投入柴油发电机出线开关。

图 2.32　室外柴油发电机

　　对于不同负荷等级的负荷供电，综合用电需求和现场实际条件，尽量利用现场已有配电柜作为区域总箱，减少对原有系统的拆改和现场设备的增加。对一级负荷中特别重要负荷由市电加柴油发电机在变配电室切换供电，弱电设备增设 UPS；一级负荷由市电加柴油发电机在变配电室切换供电；二级及三级负荷由市电供电，由于变配电室变压器设置了联络，一级、二级负荷也可实现双市电供电。供电系统单线图见图 2.33。

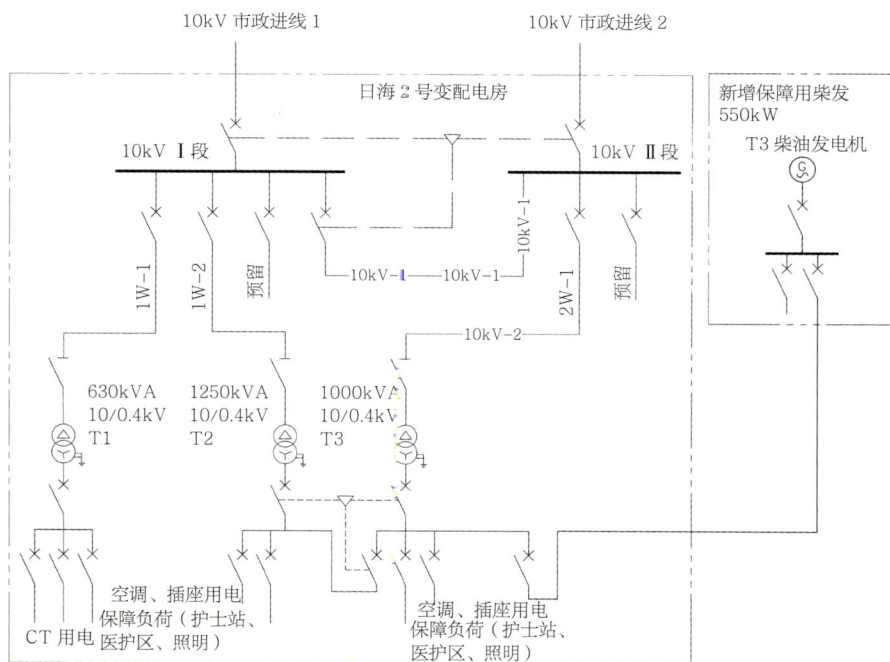
图 2.33　供电系统单线图

　　　　　　　　　　　　　　　　数字化应急医院设计及建造技术

2.3.3 低压配电系统

由于厂房面积较大且低压配电柜比较分散，低压配电系统干线主要采用放射式供电，尽量利用原有的区域配电柜作为区域总箱。

患者洗漱区主要为热水器用电，现场设置热水器专用配电箱。

隔离病房区每个病床按1.5kW配置2个单相5孔用电插座，安装高度高出每个床位床头柜0.1m为宜；在每个隔离病房区两侧各设置2个或3个单相插座，为该病区空气净化器及饮水机供电。按隔离病房区域分散设置插座配电箱，为该隔离病房区的插座供电，插座配电箱电源引自厂区原有区域配电总箱。

医护区配电箱分为照明插座箱、淋浴插座配电箱、空调配电箱、风机配电箱分别为照明插座、淋浴插座、空调和风机供电。医护区配电箱设置于医护区强电井内。

室外设备的配电主要是移动CT车（图2.34）。移动CT车仅需要总电源接入，由厂房内最近的配电箱为其供电。

改造后方舱内部图如图2.35所示。

2.3.4 照明系统

照度标准参照《建筑照明设计标准》（GB 50034—2013）的高标准值并满足其负荷密度目标值规定。医护净化区照明采用吸顶式不易积尘、易于擦拭的密闭洁净灯具[17]；在更衣室（图2.36）、淋浴间等处设置紫外线消毒灯，单独开关控制。病房隔离区选用带透明罩的荧光灯吸顶安装，且设置夜间照明。考虑供电半径不宜过长，且便于集中管理，分区域设置照明配电箱，如在医护区和病房隔离区分别设置照明配电箱。

由于货源原因，疏散照明未设置集中控制型疏散照明系统，采用非集中控制型疏散照明系统，选用自带蓄电池疏散灯具，蓄电池应急时间不低于60min。火灾时，可手动切断应急

照明配电箱的主电输出，同时控制所有非持续型的照明应急灯点亮，持续型灯具的光源由节电模式转入应急点亮模式。

2.3.5 公共广播系统

1. 设计思路及理念

方舱医院收治患者数量大，人员十分密集，日常状况下需要对人员进行信息发布及管理，紧急状态下也需要能够通知到全体工作人员及患者，故设置1套公共广播系统。公共广播系统主机设置于医护区监控室内，方便与安防系统的配合及信息的发布。

2. 医护区

医护区由集装箱板房搭建，采用吸顶式扬声器。除洁净区走廊外，在脱隔离服、脱防护服的房间设置广播，防护督导组通过视频监控系统督导医护人员脱隔离服、脱防护服的动作及顺序，对于疏漏或错误的动作，使用广播进行规范指导。

图2.34 移动CT车

图2.35 改造后方舱内部图　图2.36 更衣室照明插座布置图

3. 治疗区

中国光谷日海方舱医院由原有的厂房进行改造，治疗区为原厂房的高大空间，采用壁挂式扬声器，安装在钢柱离地2.4m处，确保高于治疗区床位顶棚，保证播音效果。

2.3.6 信息接入及网络系统

1. 信息接入

为节约现场施工时间，信息接入利用了原有厂区的电信设备，由电信服务商引出4根光缆分别进入B号、C号、D号、E号方舱，在弱电间处接入本地计算机网络。因医院有内网互通的需求，各方舱之间采用12芯光缆互联。各方舱出口带宽均为500M，满足患者和医护人员使用需求。

2. 电话系统

（1）电话信息插座的设置：医院护士站按照每5～10m²设置1个语音点的标准预留。

（2）入网方式：电话采用虚拟网接入，不设程控交换机。电话接入设备放置于各方舱监控室，运营商采用光纤引入。

3. 计算机网络

中国光谷日海方舱医院项目设计了2套网络，1套为计算机信息网络，1套为计算机安防专网；2套网络在物理上隔离。

其中，计算机信息网络通过划分VLAN网段，实现内外网逻辑隔离。内网供医护人员及医疗设备通信使用；外网承载互联网信息，共医护人员、行政人员、住院患者使用。

计算机安防专网用于承载安防监控系统的弱电通信信号。

（1）信息插座的设置。医护区计算机信息插座的设置同电话信息插座，信息插座采用双口型，一个用于语音，一个用于数据。

（2）主干网设计。主干网选用万兆以太网。在各区域监控室内设万兆以太网交换机、服务器、路由器及多媒体工作站等。传输介质采用光纤。

（3）在接入层，内外网及安防网均使用千兆交换机；方舱各区域的接入层交换机通过光纤直接接入监控室的汇聚层万兆以太网光纤交换机；汇聚层交换机通过光纤接入核心层。

（4）广域网连接。于原网络中心以专线接入互联网，电信服务商引出4根光纤至各方舱监控室，为方舱各区域提供网上浏览和电子邮件服务。

4. 无线网络

由于医院地处偏远，人员密集，无线网络使用需求巨大，故在方舱医院设置无线AP阵列，以满足移动医疗设备、患者及医护人员的无线通信及网络需求。

治疗区域、值班室、办公室、护士站、走道设置高密无线放装AP，实现无线信号全覆盖，无线AP采用POE供电。AP设置2个SSID，患者专用SSID仅提供外网接入服务，医护专用SSID同时提供内、外网接入服务。

治疗区通过管理软件限速400M，工作区限速100M，工作区取高优先级，优先满足医护人员使用。

2.3.7 综合布线系统

1. 设计思路及理念

方舱医院的综合布线系统主要满足医护办公需求。医护区设置内外网点位。

（1）信息插座。信息插座均采用6类信息插座模块。每个信息插座附近应配备电源插座。信息插座距地0.3m，于墙上明装。

（2）水平布线。水平布线采用6类4对UTP。从监控室穿线槽引出，沿走道顶板敷设至工作区后穿管或线槽明敷至信息插座。水平布线长度不超过90m，走线管槽的最大长度控制在70m。

（3）水平干线。水平干线采用3类25对

UTP 铜缆及 6 芯多模光纤。其中 3 类 25 对 UTP 铜缆用于电话通信，6 芯多模光纤用于计算机网络。干线电缆穿钢线槽在吊顶内及竖井内敷设。

（4）总配线架。电话铜缆总配线架采用 110A-300 型跳线架；数据总配线架采用光纤配线架。电话配线架及光纤总配线架安装于监控室的 19 英寸机柜内。

2. 医护区

医护区设置了外网和内网 2 套布点系统。综合布线系统信息点配置见表 2.2。

综合布线系统信息点配置　　　　　　　　表 2.2

医疗场所	标准配置
医疗检验、检查设备	每台设备设置 1 个内网数据点
检验工作台	每个工位设置 1 个语音点、1 个内网数据点
诊断报告工作台	每个工位设置 1 个语音点、1 个内网数据点
护士站	1 个语音点、3 个内网数据点
医生、护士办公室	每个工位设置 1 个语音点、1 个内网数据点、1 个外网数据点
处置室、治疗室、值班室、药品库、仪器库	1 个语音点、1 个内网数据点
公共区	无线覆盖

除此之外，D 号方舱外部设置了 3 台移动 CT 车，为满足 CT 设备接入网络的需求，在该处设置 1 台 8 口千兆交换机，移动 CT 车均由光纤接入交换机。

3. 治疗区

考虑到患者接入网络的需求，在治疗区设置了无线 AP 阵列，根据床位分布进行无线 AP 点的排列，约每 30 张床位设置 1 台无线 AP，保证无线设备负荷维持在一个合理的水平。

治疗区患者出入口设置 1 个外网语音点、2 个专网语音点，以满足新患者进入手续的管理需求，方便工作人员就地进行沟通。

2.3.8　安全防范系统

1. 设计思路及理念

方舱医院不同于火神山医院和雷神山医院，火神山医院和雷神山医院用于救治重症患者。而方舱医院收治的是轻症患者，轻症患者大部分能自主行动，且方舱医院内收治的患者数量较大，每个方舱的床位数量能达到 600 张以上，因此安全防范系统是必须要配置的系统。

2. 治疗区

各方舱治疗区按照每个单元设置 1 个监控点位，监控方舱内情况。在主要走道高处设置全景摄像机，根据现场实际情况并结合公安部门要求，安装高度在 6.0m 以上，保证能够监控方舱的整体情况，如图 2.37、图 2.38 所示。

在人员流动活动区域（休闲区、图书角、影视娱乐区），除在高处设置全景摄像机外，还加装局部的监控点位，实现从整体到局部的监控全覆盖。

图 2.37　治疗区入口设置监控

图 2.38　治疗区设置监控

在方舱的患者出入口及室外功能区设置监控点位。对于较开阔区域采用室外球形摄像机，对于较为狭长的区域（室外洗手台、室外储物柜等区域）采用室外枪机。

3. 医护区

医护区也保证视频监控无死角、全覆盖，在走道、出入口、医生办公室、治疗室等均设置监控设备。医护人员工作繁重，筋疲力尽的医护人员在脱防护服时容易有疏漏，因此在脱隔离服、脱防护服处的房间设置半球摄像机，安装在房间镜子的正上方，由后方的防护督导组监控医护人员脱隔离服、脱防护服的动作是否标准规范、顺序是否正确。对于疏漏或错误的动作，督导组通过该处的广播给予及时的提醒并采取补救措施，保障医护人员的安全。

在护士站、治疗室、医生办公室设置一键报警按钮，接入医院安全防范系统，以备出现紧急情况时由医护人员报警求助。

4. 系统组成

方舱医院采用数字监控系统，设置监控专网，与工作网络分开，保证了系统的安全性及可靠性（图 2.39）。

监控系统的接入交换机采用 1000M 局域以太网架构，在各方舱的监控中心设置 NVR（网络硬盘录像机）进行区域资料的存储、控制。

结合医院及公安部门的要求，系统应满足监控资料存储时间不少于 30 天的要求。

计算硬盘数量应考虑编码压缩算法及前端摄像机的分辨率，硬盘采用的是当时市面上能采购到的 6TB 硬盘。计算示例如下。

E 号方舱医院：90 路存储 30 天的 1080P（200 万像素）视频格式的录像信息的存储空间大小为 512（码流，kb/s）×3600（s）×24（h）×30（天）×90（路）÷0.9，即 126562.5GB，约为 123.6 TB，需要 21 块 6TB 硬盘，设计配置 24 块 6TB 硬盘。

图 2.39　监控室调试现场

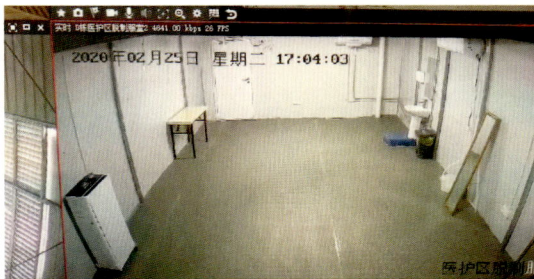
图 2.40　医护区实时监控画面

图 2.41　室外实时监控画面

在指挥中心设置安全防范监控平台，各方舱的 NVR 接入监控平台，并通过拼接大屏进行重点区域的实时监控显示（图 2.40、图 2.41）。

考虑到方舱医院人员密集，公安部门为避免出现紧急情况，要求方舱医院监控接入公安专网。公安专网的光纤通过核心交换机接入方舱医院安防系统，安防系统接入公安监控平台，满足公安部门远程监控的需求。

2.3.9　出入口控制系统

方舱医院的出入口控制系统主要设置在医护区，保证授权人自由出入、限制未授权人进入未获授权区域、对强行闯入的行为进行报警，从而保证医护区的安全。

出入口控制系统可以根据医院的运行管理需求，对医院的出入人员进行管理，确保医院安全、有序运行，并应与火灾自动报警系统联动，宜与视频安全防范监控系统联动。

当发生火灾时，门禁系统接收到消防报警系统发出的信号，对所有的门进行解锁。当门禁系统正常开门时，报警系统撤防，工作人员可以自由工作；当门禁系统非正常开门时，报警系统布防，将报警图像在监控中心的工作站上显示出来，并进行录像。

医护人员工作非常繁忙，为避免他们频繁刷卡，不宜大面积设置门禁点位。在对外出口处设置进刷卡、出按钮的模式，而为了避免误闯误入，在进入污染区、治疗区的门上设置双向刷卡的模式。可以根据方舱医院的工作管理需求，通过对卡片设置不同的权限，协助医院进行人员及流程管理。

系统采用非接触式卡片。经过与施工单位沟通，非接触式卡片市面上货源充足，易于采购，技术成熟且便于施工，能够满足现场进度要求。

2.3.10　总结

在抗击新冠肺炎疫情期间，方舱医院作为临时性集中隔离、收治轻症患者的医院，在应收尽收方面发挥了巨大的作用。方舱医院往往需要在最短的时间内建成，从设计到施工完成，工期往往只有不到 5 天的时间。由厂房改造为方舱医院，存在很多局限因素，如原有供电条件不好、原有设施不齐备、环境较差等，而且方舱医院又受到诸多医院设计规范的限制，需要做到安全合规、系统简洁、便于施工，在尽力满足规范条文要求的情况下，结合实际情况进行合理的创新和改进。

通过对中国光谷日海方舱医院的改造，笔者总结出在将厂房改造为方舱医院的过程中，应做到以下几点。

（1）注重现场踏勘，了解现场电气设施的安装使用情况，结合改造需求制定合理的电气改造方案，如在满足使用功能要求的前提下尽量利用原有的配电总箱、照明系统、火灾报警系统等，减少设备的拆改或增加，缩短工期。

（2）电气系统简洁、可靠、一致，比如多个方舱的设计原理保持一致，设计快速，施工单位也能迅速了解设计理念，准确快速施工。

（3）充分了解方舱医院的使用维护需求，比如病床插座的使用需求。中国光谷日海方舱医院中隔离区插座主要用于满足电热毯、台灯、饮水机、带杀毒功能的净化器等设备的用电需求。只有充分了解需求，才能做到不遗漏，避免施工返工。

（4）所选用电设备应满足简单可靠、施工快速、调试方便、货源充足的要求，并时刻与施工单位保持密切联系以便及时解决问题。比如设计中采用集中控制型疏散照明系统和 48 盘位的 NVR，但由于工期紧张，且特殊时期各种货源并不充足，后经过与施工单位、厂家沟通，将集中控制型疏散照明系统改为非集中控制型疏散照明系统，采用 2 台 24 盘位的 NVR 替代 48 盘位的 NVR，后经施工调试，系统均运行正常。

2.4　通风空调改造技术

自新冠肺炎爆发以来，为了隔离病毒传播，政府在极短时间新建多所方舱医院，主要收治轻症确诊患者。所建方舱医院绝大多数由既有高大空间（如体育馆、展厅、工厂等）改造而成，当用作新冠肺炎患者治疗场所时，为保证医护人员、工作人员、患者的健康、防止病毒向周围环境扩散，如何在极短时间内进行改造是值得探讨的问题。

笔者在参与多座方舱医院改造的过程中，发现清洁区的改造简易可行，但是如何合理地进行污染区、半污染区的改造应慎重思考，以下进行具体分析探讨。

2.4.1 改造建筑特征

1. 建筑特征

武汉市用于改造为方舱医院的既有建筑有体育馆、展厅、工厂等，它们具有一些共同特征。①建筑周边空间较大，交通便利，距离居民区较远，水、电、通信等设施齐全。②建筑空间高大，一般室内高度大于15m，厂房高度稍低，但是室内高度也大于5m；门窗多，屋面大多数为网架结构，气密性较差。③建筑均设有集中空调系统。④消防设施基本齐全。

2. 通风空调及排烟设施

体育馆、展厅空调末端方式为一次回风全空气空调系统，采用喷口中部侧送风，下部集中回风。部分体育馆的场馆、展厅设有机械排风和机械排烟系统。机械排风系统的排风量小于空调送风量。除空调送风方式采用顶送风外，厂房的通风、空调及排烟形式与体育馆、展厅基本相同。所有排风口高度均不高于屋面2m，且排风口与新风口的相对位置比较随机。

3. 通风空调改造要求及特点

根据职能部门、使用单位及相关规范的要求，方舱医院的改造应达到以下目标：①改造后的方舱医院能够保证医护人员、工作人员、患

者的健康，防止污染区、半污染区的病毒向周围环境扩散；②由于用于改造的时间极短，通风空调改造需基于现状，所需设备、材料应尽量利用现有物质或便于采购；③方舱医院使用完毕后能够迅速、简单地恢复原有功能。

2.4.2 方舱医院通风空调系统改造分析

方舱医院污染区、半污染区的基本功能如图2.42所示，以下就每个区域的通风空调改造进行分析。

1. 病区

病区主要功能包括病床区、护士站、抢救治疗室、活动室、备餐间、被服库等，轻症患者按病情分隔在多个区域，隔断高度一般为2m左右，病人人均使用面积为7～12m²。容纳1000人的方舱医院，每天医护人员分为4班，每班人数约为60人。

（1）病床的布置

许多呼吸道传染性疾病（包括新冠肺炎）爆发于冬、春季，此时室外温度较低，室内需要供热，靠近外窗区域会形成贴壁气流。室内外温差越大、窗户隔热性能越差，贴壁气流越强，当处于该区域的人互相交流时，贴壁气流会破坏人体表面的热羽流[18]，易形成交叉感染，因此病床布置应远离该区域，同时也能避免冷辐射对病人的影响。由于病床区人员较为密集，可以设置医用消毒净化装置以降低各种疾病的感染风险。

（2）高大空间送排风量对室内负压的影响

清洁区、半污染区、污染区需要保持有序的压力梯度，病区的压力最低，但是高大空间的气密性较差，如何保持合理的压差，防止室内空气外溢值得探讨，以下以展厅、体育馆、厂房改造为可容纳1000人的方舱医院为例进行分析。

展厅：方舱医院位于建筑的一层内区，层高15m。体育馆：方舱医院位于一层，层高30m，一层为内区，二层上部有外窗和网架结构屋顶。

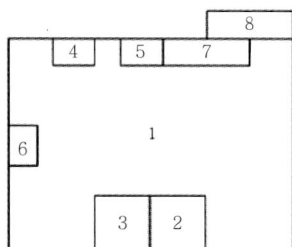

1—病区；2—医护入口；3—医护出口；4—病人出口；5—污物出口；6—病人入口；7—辅助用房；8—室外厕所

图2.42　方舱医院功能示意图

厂房：方舱医院位于建筑的一层，层高6.6m，有外窗。

病区的排风量标准按每人不小于150m³/h计算[19]，排风量为150000m³/h。在门窗缝隙边发烟测试气流效果表明，当送风量为排风量的90%时，无外窗和网架结构屋顶的展厅改造成的方舱医院能保证负压状态，有效防止病毒向周围环境扩散，但是由于无外窗，患者生活环境的体验感较差，压力大，不利于轻症患者恢复健康。体育馆由于外窗较多且有网架结构屋顶，改造成的方舱医院部分区域不能保证负压状态，当送风量变为排风量的60%时，室内负压情况有所改善，其主要原因是体育馆空间太大，漏风量较大，排风系统布置不均匀。厂房层高较小，有外窗，送风量相对排风量的大小及室内负压情况处于以上两种情况之间。通过以上分析可以得出，为避免病毒外溢，选择高度合适、有较少外窗的厂房或展厅较为合理。当建筑密闭性较差时应减少送风量以维持室内负压，送风量的大小建议按排风量的60%~90%取值，密闭性较差时取低值，反之取高值。

（3）通风空调改造方式

改造末端空调直风系统有两种方案：①全新风运行模式；②新风回风混合运行模式，二者的优缺点见表2.3。

从表2.3可以看出，采用全新风运行模式更适合疫情时的改造方案。

（4）全新风运行模式室内温度分析

方舱医院实际运行情况表明病区内的温度对病人的情绪影响非常大，而新冠肺炎爆发于冬、春季，该时期属于供暖期，当采用全新风运行模式时，室内温度影响程度分析如下。

对于长江中下游地区，建筑的空调热负荷一般为冷负荷的40%～60%，室内以空调冷负荷为基准选取空调器，通过计算得出，在整个供热系统所有区域均运行的情况下，冷热水共用盘管的空调器在全新风运行时，送风温度在室外温度的基础上提升约11℃。当部分区域不供热运行，仅改造的方舱医院区域运行且热源的容量满足方舱医院全新风运行时所需的热负荷，此时供热能力取决于空调器自身供热能力和热水流量。热水流量按照夏季冷水流量取值，通过计算得出：供热水温为60/50℃，六排管的空调器在全新风运行时，单位风量（m³/h）所提供的热量约为10~12W，若按12W取值则送风温度在室外温度的基础上提升约33.6℃。若

全新风运行模式与新风回风混合运行模式的优缺点　　　　　　　　　　　　　　　　　　　　　　　　表2.3

内容	全新风运行模式	新风回风混合运行模式
改造方案	①关闭回风阀，全开新风阀；②增设排风过滤器组（或其他有效消毒措施）和排风机（若无其他有效消毒措施）；③调节新风量保持室内为负压	①增设回风高效过滤；②调节新风、回风阀门；③增设排风过滤器组（或其他有效消毒措施）和排风机（若无其他有效消毒措施）；④调节新风量保持室内为负压
存在问题	①新风未有达到亚高效级过滤级别，若增加则空调器的风机压力不够；②由于平时空调区保持正压要求，排风机的风量小于空调器送风机的风量	①回风口需尽量扩大面积，安装高效过滤器以降低阻力；②新风没有达到亚高效级过滤级别，若增加则空调器的风机压力不够；③回风口高效过滤器运行时阻力变化较大，难以调节新风、回风比例，导致室内负压难以保证
建议	①空调器预留亚高效过滤段，空调器风机选用性能曲线平缓型以保证疫情时送风；②排风机选用变频方式，平时变频，疫情时工频运行，空调器采用变频方式，平时根据需求采用工频或变频，疫情时采用变频方式	①空调器预留亚高效过滤段，空调器风机选用性能曲线平缓型以保证疫情时送风；②排风机选用变频方式，平时变频，疫情时工频运行
室内换气次数	高	低
室内温度保证	较难	较易
改造难易度	较简单	较复杂
改造投资	较小	较高
运营维护	工作量较小、感染风险小	工作量较大、感染风险高

要提升送风温度，则可以采取以下两种措施：①加大热水流量，但其受限于既有空调管径及热水循环水泵的扬程，流量增加的幅度很小；②提高供水温度达到65℃时，供热能力提升约12%，即送风温度可以提升约37.6℃。由此可见，正常情况下，当回风状态的空调器改成全新风运行模式时，其送风温度在室外温度的基础上提升11~33.6℃。

若病区的排风量按每人不小于100m³/h计算，空调器的新风量按排风量的80%取值，人均占地面积10m²，当室外温度为-2℃时，为满足室内温度达到16℃的要求，根据上面分析，空调器最高提供的热负荷约为47W/m²，渗透风形成的热负荷约为13W/m²，剩余的热负荷为34W/m²，基本满足维护结构所需的热负荷的要求。

由上得出，当室外温度为-2℃时，正常供热系统全部开启，方舱医院采用全新风模式运行，室内温度不能达到16℃。当供热系统采用60℃的供水温度，病区按患者每人不小于100m³/h的排风量进行排风，关闭除方舱医院使用区域外的其他供热末端，空调热源能够满足方舱医院病区的热负荷需求，采用六排管的空调器能够达到室内温度为16℃。当排风量增加时，方舱医院区域的热负荷需求加大，热源不能满足病区的热负荷需求时其室内温度变低。当空调器采用四排管时，其送风温度低于六排管空调器的送风温度，即单位风量所提供的热负荷减少，为了维持16℃室温则需加大人均送风量和排风量，病区所需的供热负荷更大，室内温度更难以保证。

严寒地区采用直流式空调系统时，若室内采用其他供热方式如地板辐射供暖，则较容易达到16℃的室内温度。寒冷地区室内无其他供热设施，室外温度低，则室内无法达到16℃，需要采用其他辅助措施如电热器以提高室内温度，从而保护病人健康及情绪。

（5）人均排风量分析

病区的排风量取值多少值得探究，其大小需满足以下条件：①满足病人新风量需求；②建立有序的压力梯度；③保证室内合适的温度；④有利于迅速改造完成。

由于方舱医院收治的是确诊轻症患者，病人所需的新风量应不小于40m³/h[20]。高大空间改为方舱医院，其密闭性较差，室内难以达到-5Pa的绝对压力，只能与室外保持相对的压差以防止病毒向室外扩散，降低周围区域感染风险。根据上面的分析，当排风量按每人不小于100m³/h取值时可以满足室内16℃的温度，加大排风量则很难保证室内温度。排风量增大意味着排风高效过滤器组也相应增大，改造难度增加，施工时间变长，投资也增大，作为临时使用不太可靠、经济。

按照《传染病医院建筑设计规范》（GB 50849—2014）规定，呼吸道传染病的病房最小新风换气次数为6次/h。若病房按人均10m²、层高2.4m计算，6次/h的排风量为144m³/h。人体平和状态下呼吸空气量约为6~9L/min，即0.36~0.54m³/h，由于方舱医院长时间收治患者，其病毒产生率很高，当患者停留160min时，达到4680quanta/h，由图2.43[21]得出，采用10次/h以下的换气次数不能有效地降低感染风险。

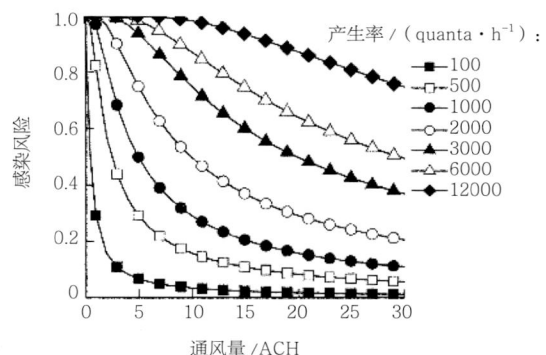

图2.43　感染风险、quanta产生率及通风量变化的关系

来源：钱华，郑晓红，张学军．呼吸道传染病空气传播的感染概率的预测模型 [J]．东南大学学报，2012，42（3）：468-472.

为满足患者的新风和室温需求，保持室内有序的压力梯度和改造的易行性，建议排风量按每人不小于90m³/h取值。对于夏热冬暖地区可以适当提高排风量。

综合以上分析，结合体育馆、展厅和厂房的空调通风设置情况，以空调通风系统为基准，厂房和展厅更适合改造为方舱医院，体育馆稍差。

2. 医护出入口

医护出入口是防止交叉感染的重点关注区域，该区域均由体育馆、展厅及厂房的附属房间改造而成，根据"三区两通道"要求，典型医护出入口和通风示意如图2.44、图2.45所示。

（1）医护人员入口

医护人员进入需经过一更、二更、缓冲间到达病区，由于病区为负压，一更、二更采用送风方式保证正压、防止病毒扩散至清洁区是合理的。工程中采用压差法计算门窗缝隙泄漏风量，其简化公式为

$$L=0.827 \times A \times \Delta P^{\frac{1}{2}} \times 1.25$$

当一更送风时，各相邻隔间采用DN300的通风短管，按单扇门（2m×0.9m，门缝4mm）、压力为10Pa计算（病区按绝对负压 –5Pa计算，较难达到），外门、内门缝和短管的泄漏风量分别为193m³/h、272m³/h、828m³/h，则总送风量必须保证大于1293m³/h。当一更采用30次/h换气进行送风时，其体积必须大于43.1m³（净高2.4m，面积18m²）才能保证与病区有10Pa压差，因此设计中必须按一更30次/h换气和1293m³/h取大值选择送风机。对于可容纳1000人的方舱医院，每班医护人数约为60人，建议一更的面积为20m²以上。

为了便于改造、快速安装，同时考虑运行时送风过滤器的阻力越来越大，建议送风机选择平缓型性能曲线的定频风机；二更与缓冲间的短管上设置手动密闭风阀，以便风机产生故障时进行安全防护；缓冲间与病区采用带电动风阀的短管，与送风机连锁；为防止门自然开启泄压，其开启方向也需随压力梯度确定。

（2）医护出口

医护人员离开病区一般需经过缓冲间、脱隔离服间、脱防护服间、脱制服（淋浴）间、更衣返回清洁区。各空间使用特点为：①缓冲间医护人员通过时间极短；②脱隔离服房间病菌浓度可能最高，医护人员需经过脱手套、护目镜、隔离服、面罩、鞋套、洗手等过程，停留时间较长，防护服及口罩有可能沾上病毒；③冬季脱制服（淋浴）间热气较大。改造要求为：保证有序的压力梯度，气流合理流动，工程施工简单易行。

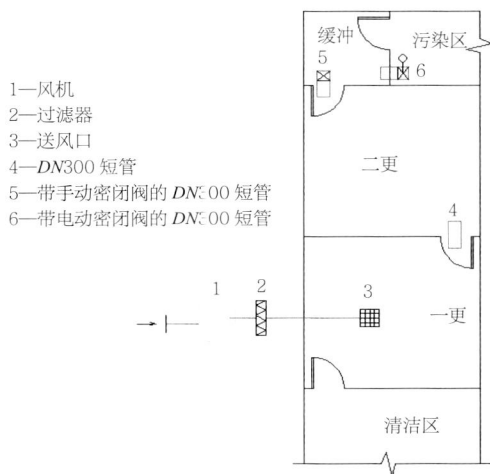

1—风机
2—过滤器
3—送风口
4—DN300短管
5—带手动密闭阀的DN300短管
6—带电动密闭阀的DN300短管

图2.44 入口及通风示意

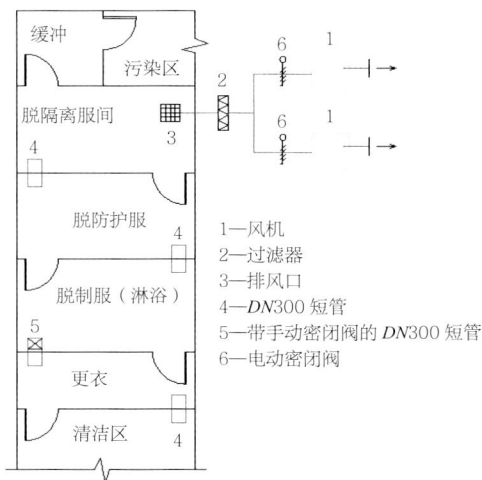

1—风机
2—过滤器
3—排风口
4—DN300短管
5—带手动密闭阀的DN300短管
6—电动密闭阀

图2.45 出口及通风示意

根据排风方案可以建立如下压力梯度：脱制服（淋浴）间 -5Pa、脱防护服间 -10Pa、脱隔离服间 -15Pa、病区 -5Pa，在脱隔离服间设置换气次数不小于 30 次 /h 的排风设施、其他相邻隔间采用 DN300 的短管是简单易行的方式。

缓冲间的通风方式值得分析。其两侧的病区及脱隔离服间均为污染区，若采用送风方式可阻止病区病毒流向脱隔离服间，但是其送风量很小，系统复杂，施工难度大；若仅在脱隔离服间设置排风，则缓冲间通过门缝渗透的风量为 193m³/h，当其面积小于 10m²（净高 2.4m）时，室内换气次数达到 8 次 /h，因此可以不设置排风，同时缓冲间的面积不宜过大。

为保持脱隔离服间的负压，同理计算其排风量必须保证大于 976m³/h；采用 30 次 /h 换气计算时，其体积必须大于 32.5m³（净高 2.4m，面积 13.5m²）才能保证与病区的压差，因此设计中必须从脱隔离服间换气次数 30 次 /h 和 976m³/h 二者中取大值选择排风机。可容纳 1000 人的方舱医院，每班医护人数约为 60 人，建议脱隔离服间的合适面积为 25m² 以上，采用 30 次 /h 换气能够保证有序的压力梯度。

医护人员在脱隔离服间的停留时间约为 5min，且室内气流组织不一定均匀，因此防护服及口罩有二次污染的可能性，建议在脱隔离服间、脱防护服间增设风淋室（顶出风），可以有效地排出医护人员身上所吸附的病毒。冬季脱制服（淋浴）间热气较大，排风机不一定有效排出热气，可根据实际情况在此增设排风系统或加大脱隔离服间的排风机风量。脱制服（淋浴）间与脱防护服间采用带手动密闭阀的短管是为了在排风机因意外事故停止运行时起隔绝作用，排风机也可设置备用风机。为防止门自然开启泄压，其开启方向也需随压力梯度确定。

（3）患者出入口

患者入口包括寄存、消毒、安检、更衣等空间，患者出口包括转院和康复出口，均设置消毒和打包区域。该区域需设置排风以满足两个要求：①防止病毒通过外门逸出室外；②由于出入口经常进行消毒，需要排除消毒剂所产生的异味。对于患者入口区，其负压值可以与病区相同，当维持 -5Pa 的压差时，按 1 个室外单扇门计算则排风量为 193m³/h；若按换气次数采用 12 次 /h 计算时，其排风量一般远大于 193m³/h；因此入口区的负压会低于病区负压，可以作为安全屏障。

对于患者出口区，由于转院病人可能需要病床，双扇外门开启时间较长，压差并不能完全保证。该区域为负压且负压值可以与病区相同，当维持 -5Pa 的压差时，按 1 个室外双扇门计算则排风量为 320m³/h；由于开门时间长，因此建议按换气次数 12 次 /h 计算，且取大值。

（4）辅助用房

病区内还包含辅助用房，如抢救治疗室、处置室、库房、开水间、污洗间、垃圾间、污物通道等，其中抢救治疗室、处置室等采用半隔断空间，因此不需要特别处理。若为封闭空间则需要增设通风系统，当进风有条件从室外取风时，则需要增设电动风阀以保证风机出现故障时关闭新风口；当无条件时则直接从病区取风；排风可以直接排至病区。

开水间、污洗间、垃圾间、污物通道需要设置机械排风系统，排风量可以按照 12 次 /h 的换气次数计算取值。

3. 室外卫生间

由于患者使用的卫生间的排水需要特别处理，本次改造的方舱医院的卫生间全部采用成品卫生间且在室外设置，从使用效果看，并未对周围工作人员产生影响。本着简单易行且平灾结合的原则，卫生间建议设置在灾区灾时高频风向的下风侧。根据大气扩散模型[22]中大气扩散浓度估算，室外风速为 2m/s 时，100m、200m 距离的浓度稀释约为 450、1500 倍，因此建议最好周边 150m 范围内不能有居民区，此时浓度稀释约为 1000 倍。

4. 其他

以上所有排风机均需要设置初、中、高效过滤器组；排风口的位置最好位于高处且下风侧。新风口与排风口之间应保证 20m 的水平距离，当不满足条件时，可以在垂直方向上设 6m 以上高差。

2.4.3 设计体会

1. 通风物资储备

新冠肺炎疫情暴发时，患者的数量快速增长，为了在最短时间内建设好方舱医院，应对今后可能发生的类似的呼吸道传染性疾病，所有有关通风系统的物资需要进行灾备储备，主要包括风机、过滤器、风淋室、医用消毒净化装置、空气过滤器、压差报警器等。建议所有病区的设备以患者数 200、400 为模数对应配置送风机、亚高效过滤器组（H11，包括粗效、中效过滤器）、排风机、高效过滤器组（H13，包括粗效、中效过滤器）；医护人员、患者出入口及其他区域的风机也作相应配置。

2. 预留氧气管道

由于患呼吸道传染性疾病的患者需要大量氧气，因此对于适宜改造为方舱医院的公共建筑需预敷设氧气管道，地面预留氧气快速接头。当灾害发生时可以快速对接使用，能够避免氧气瓶组进入方舱医院所引起的危险，也减少更换氧气瓶组所带来的感染风险。

3. 建筑设计

对于某些公共建筑，政府应制定相应的法规，设计时应考虑平灾结合方式，为今后改造方舱医院创造有利条件，特别是医护人员出入

口所需的功能条件、预留风机所需要的配电回路，等等。

4. 方舱医院通风空调系统改造建议

（1）通风空调系统改造以切断污染传播途径，避免交叉感染及对周边环境影响为原则。

（2）采用机械通风方式保证清洁区、半污染区、污染区有序的压力梯度且空气静压依次降低，清洁区为正压，半污染区、污染区为负压；各分区需分别设置独立的通风系统。

（3）排风机应设置在排风系统末端且位于室外，宜设备用排风机，排风需经过高效过滤后高空排放。送风系统宜经过亚高效过滤后送入室内，过滤器宜设置压差报警装置。

（4）医护入口宜采用正压控制方式，医护出口宜采用负压控制方式，患者出入口、物流通道等应采用负压控制方式。

（5）室内应保证良好的气流组织，局部区域可根据使用要求设置医用空气净化器。

（6）送风系统和排风系统宜设连锁控制，启停顺序根据室内静压要求确定。

（7）排风口应设置在疫情期间主导风向的下风侧，新风口与排风口须保持合适的距离，以避免互相影响。

（8）室内温度宜为 14~28℃，室内温度不能保证时可采用其他临时辅助设施。

（9）采用全空气系统的病区通风空调系统宜改造为全新风运行模式，当采用回风模式时须采用可靠的回风高效过滤措施。

（10）采用冷热末端 + 新风系统的病区宜把新风系统改造为排风系统，通过外窗自然补风，室内须保持负压，排风系统须保证不间断运行。

（11）半污染区、污染区的空调凝结水管须保证水封可靠，凝结水应集中收集并排至指定点。

（12）排风高效空气过滤器更换须由专业人员进行消毒灭菌，并随医疗废弃物一起处理。

（13）病区根据使用需求配置移动式氧气瓶。

1

Leishenshan Hospital

1.1 Application of Digital Technology in Construction Industry

To help people in distress as quickly as possible, Central-South Architectural Design Institute Co., Ltd. (CSADI) completed the construction task at a lightning speed within just about 10 days from January 24, when CSADI received the design task of Leishenshan Hospital, to February 8 when Leishenshan Hospital was officially put into use. What CSADI did played an important role in curbing the spread of COVID-19 epidemic. Apparently, the exquisite and efficient digital technology made a difference in the construction of Leishenshan Hospital and even in the field of architectural design.

1.1.1 Necessity of Digitalization in Construction Industry

In today's world, scientific and technological revolution and industrial transformation are progressing with each passing day. The booming digital economy has profoundly changed the mode of human production and life, and has a far-reaching influence in the global economic and social development, the global governance system, and the process of human civilization. Chinese manufacturing, creation, and construction make astonishing achievements and keep changing the outlook of China. Digital technology, backed up by the digital economy, is constantly integrated with the architectural design industry to generate sustained "chemical reactions". This profoundly changes the production mode of construction industry, and lays a solid foundation for upgrading the construction industry to a higher level for quality development.

Digitalization enables the interaction between business and technologies, changes traditional business operation modes, and creates new scopes of business and new opportunities for creating income and value.

The concept of digital technology is of broad sense. Regarding the construction industry, it focuses on the BIM technology and integrates various information technologies, such as parameterization, virtual simulation, 3D visualization, internet of things, cloud computing, mobile Internet, big data, and artificial intelligence. Relying on various softwares, it completes the digital delivery and storage of building models, realizes the "digital twin" of digital virtual pattern and architectural entity for construction products, effectively and rapidly finishes design, simulation and optimization, generates better solutions, and provides overall guidance on the implementation of architectural design, so as to improve efficiency and reduce costs.

In the context of global digital economy, we are facing high pollution, high energy consumption, and low efficiency of Chinese construction industry. For us, digitalization is not an option but the only way out. We have to make use of digital means to establish project-based digital systems for production and operations, extend the application range of digital technology based on enterprise-level information systems, and improve the application effect of digital technology from conception to building O&M (operation and maintenance). In addition, we also need to

carry out in-depth R&D (research and development) in digital generation, analysis and optimization of digital technology, management of digital technology, and establishment of the ecological system. In addition, we also have to complete the R&D for the demand matching of "the last kilometer" for the unified business platform, to realize the digital transformation and upgrading in production, organization, and value chain of construction enterprises.

1.1.2　Characteristics of Digital Application in Construction Industry

In early days of the computer age, some practitioners completed complicated mechanical analysis and calculation via computer programs after digitalizing structural models of buildings. However, the rapid development and popularization of digital technology had not affected the construction industry until the large-scale advanced manufacturing industry represented by the aviation industry realized digitalization (Figure 1.1).

On one hand, as the early digital construction requires huge investments, only the aviation industry can afford the cost. On the other hand, compared with digital models from the manufacturing industry, the one of the construction industry features massive data and large span on sizes. Therefore, the needs of the construction industry can be met only when the digital software and hardware technologies are developed to a certain level.

Meanwhile, the construction industry and advanced manufacturing industry also share many similar characteristics. Taking aircraft

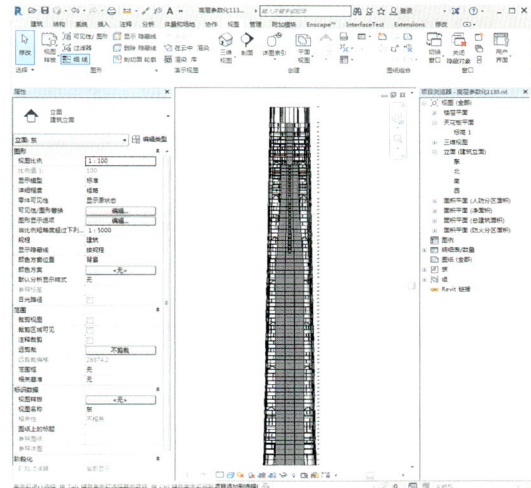

（a） （b）

Figure 1.1　Transfer of Digital Technology from Advanced Manufacturing to Construction Industry
（a）Large advanced manufacturing industry；（b）Digitalization of construction industry

manufacturing as an example, the aircraft manufacturing is characterized by large data volume, numerous changes and task categories, strict manufacturing cost control, long supply chain, and plenty of suppliers. While for complicated construction projects, they are also similarly characterized by large quantity of information, numerous design and construction modifications, many disciplines involved, tight control of construction cost, long project duration, and cooperation with plenty of subcontractors, which enable the digital experience gained from large manufacturing industry to be technically transferred to the construction engineering industry.

The application of digital technology in the construction industry can be illustrated from the following three aspects. First, digital technology can be used to establish the geometric models of buildings and surrounding environment, store the building lifecycle information, set up the business processes for the full li-

fecycle of buildings, and complete the digital expression of buildings in a more efficient way. Second, digital technology can realize efficient generation, accurate analysis, and quality management of the construction industry aided by efficient computer systems. Finally, digital technology help reshape the business model of the construction industry through technological innovation, thereby accelerating and empowering the future development of the construction industry.

1.1.3　Digital Expression of Construction Industry

For digital expression of buildings, digitalization and visualization of geometric shapes and structural composition is prioritized. From hand-drawn drawings, 2D CAD, to increasingly prevailing BIM models, or even virtual reality, the digital expression of buildings becomes increasingly adaptive to human observation habits（Figure 1.2）.

Figure 1.2　3D BIM Model of Buildings

Figure 1.3　Virtual Reality（VR）

The virtual reality（VR）and augmented reality（AR）technologies go further in experience. They provide users with realistic visual perception by simulating digital buildings in virtual space with computers. Users can immerse themselves in a simulated environment, obtain the same or similar perception as the real world, and feel a strong sense of presence. Besides providing the experience of real space, the VR and AR technologies can also display the digital information invisible in the real world, such as light, heat, energy, wind, and electric field by virtue of science and technology, thus helping users enhance cognition, understand concepts, cultivate creativity, and inspire productive thinking（Figure 1.3）.

Upon the 3D digitalization of buildings, engineering measurement is just a piece of cake. By extracting information from the 3D digital models of a building, we can obtain any property of the building in terms of the stage, time point, system, floor, and discipline, and quickly extract such information for use, without worrying about which floor, which section, vertical or horizontal component, and what specifications and models to extract.

Besides the architectural discipline, the construction industry covers many other disciplines, such as structure, water supply and drainage, HVAC（heating, ventilation and air conditioning）, and electrical disciplines, and involves the knowledge of physics（mechanics, thermology, electromagnetism, optics, acoustics, etc.）, chemistry, biology, geography, and geophysics, etc. On the basis of 3D building models, the construction industry realizes digital expression of multi-specialty and multi-disciplinary knowledge through digital technology.

Engineers can extract the information related to their own disciplines from the digital building models, carry out various analysis, simulation, and optimization, adjust the design proposal of project according to the calculation results of simulation, and make furthercalculations for the new proposal until a satisfactory design proposal is obtained. Typical application scenarios include analysis of vibration mode for highrise building structures, elasto-plastic dynamic analysis of highrise shear walls, seismic response analysis of frame structures, tunnel excavation, construction impact on environment, and calculation of shear zone for slope stability. Digital expression of CAE analysis results is as shown in Figure 1.4.

Digital technology, as extending to the downstream stage of the construction industry and entering the construction or even operation stage, can create values for the full lifecycle of the construction project. Building lifecycle management (BLM) covers the full lifecycle of a construction project from planning, design, construction, O&M, to scrapping and dismantling (Figure 1.5). The full lifecycle management of buildings is extremely critical, since construction projects feature large investment, high technical content, long construction duration, high risks, and numerous parties involved.

The concept of full lifecycle management for buildings requires all parties to involve in the construction project. In the design stage, engineers must consider the buildability of the construction stage through construction simulation, and provide convenience and economic benefits for building O&M.

Construction simulation technologies make virtual verification and optimization of engineering construction proposals a reality, and can realize construction process verifica-

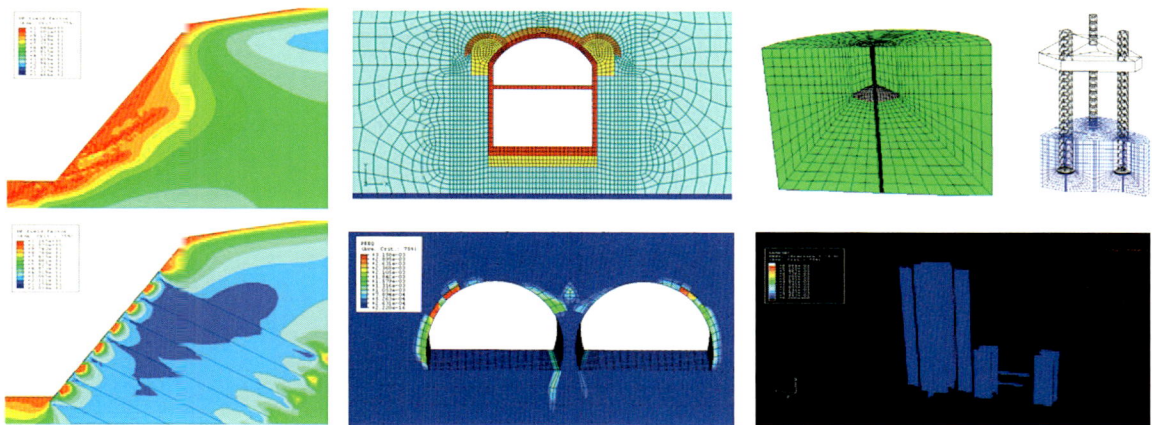

Figure 1.4   Digital Expression of CAE Analysis Results

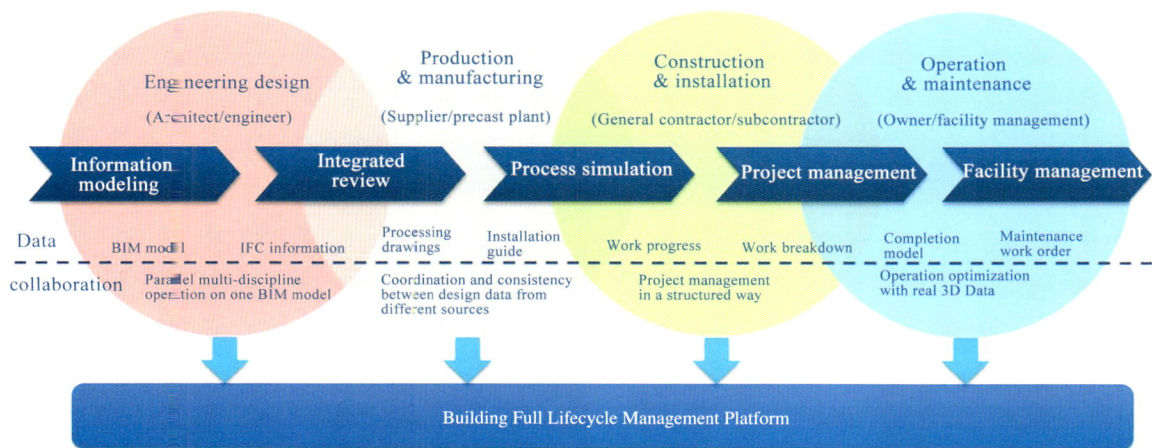

Figure 1.5   Building Full Lifecycle Management Platform

Figure 1.6   5D Simulation of BIM

tion, construction schedule verification, key process verification, and motion interference inspection, for the purpose of avoiding mistakes, omissions, collisions, and defects, and eliminating various conflicts and risks in the construction process. By comparing and analyzing the feasibility of different construction proposals, engineers can effectively control the construction cost and minimize safety accidents.

On the basis of a 4D model built by 3D space and time, construction simulation can also add another dimension of cost to establish a 5D mod-

el, which can refine budget and visualize project cost. Through 5D simulation of engineering project, engineers can obtain the accurate quantities of all building components, and statistically analyse man-hour and cost, thus controlling the construction cost (Figure 1.6).

Upon completion of a building, the building full lifecycle management platform will complete digital delivery from the construction contractor to the owner or operator. Then, the building enters the O&M stage.

In the O&M stage, the building full lifecycle management platform provides such functions as maintenance and repair instructions, allocation of maintenance and repair tasks, and inquiry of equipment parameters. Based on the IoT (the internet of things) technology, the management platform visually presents the data deployed and collected by the sensor, such as status information of equipment, pipeline pressure, $PM_{2.5}$ data, temperature and humidity,

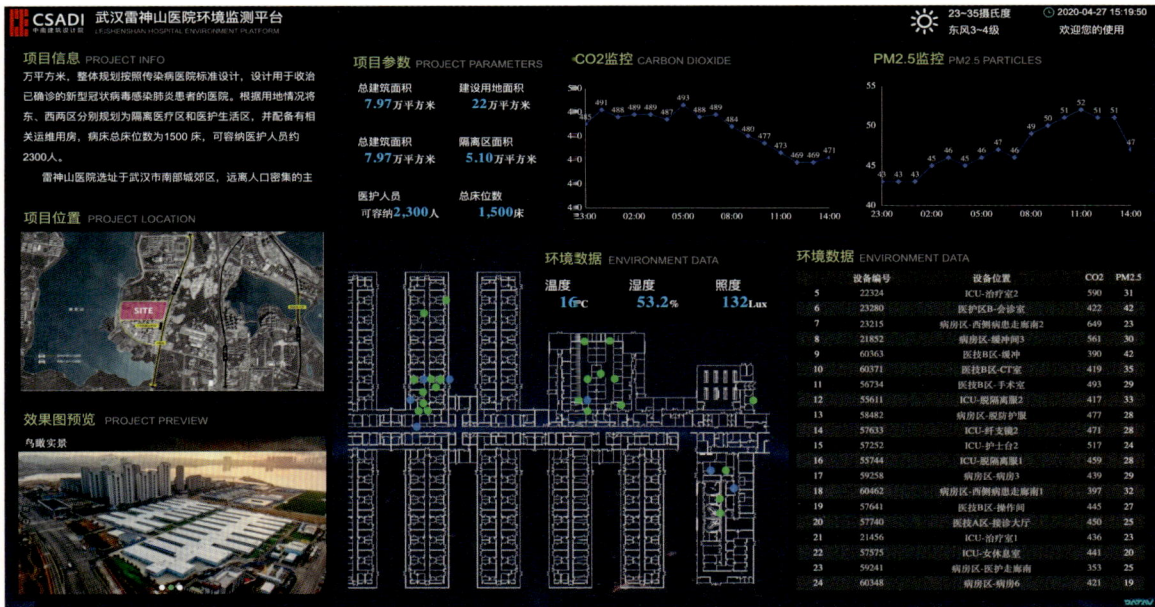

Figure 1.7   Building Operation, Maintenance and Management

and energy consumption (Figure 1.7) .

In addition to supporting the building itself, digital technology can also digitize roads, adjacent structures, etc. around the building to establish a local digital model for the building and its surroundings. When all buildings in a city are digitized, the digital model is expanded from a BIM model to a CIM model, namely the city information model.

Moreover, digital technology also helps build a twin city in the virtual space by simulating the physical world, and make scientific decisions on city operation and governance based on the digital city model. The virtual Singapore is such an example. The National Research Foundation (NRF) of Singapore, a department of the Prime Minister's Office, utilized topographical data and real-time dynamic data to establish a 3D digital twin city for Singapore so that urban planners can simulate the testing on innovative solutions in a virtual environment (Figure 1.8) . Besides the traditional map data and terrain data, the twin city also incorporates real-time data on traffic, population, mobile communications, schools, health services, energy, real estates and climate, which helps urban planners conduct virtual experiments. For example, fluid dynamics simulation can simulate the airflow around buildings, streets and green spaces, and plan the barrier-free access for the disabled and the elderly in architectural design.

Digital technology can establish a digital terrain model to simulate the physical geographical environment and the actual construction site by importing data including survey points, LiDAR, point clouds and tilt photography, and generating the terrain grid plane (Figure 1.9) . In this way, it can complete the automatic update of contour data, elevation coloring, and simulation of watershed and surface runoff, site layout and earthwork planning.

1.1.4　Quality and Efficiency Improvements of Digital Technology in Construction Industry

Digital technology will realize the catalytic enablement of the technological development in the construction industry on the basis of the digital expression. It includes efficient generation, precise analysis and high-quality management of digital technology.

I. Efficient Generation of Digital Technology

Parametric design tools, supported by digital technology can automatically finish complex operations such as the selection and combination of assembled building compo-

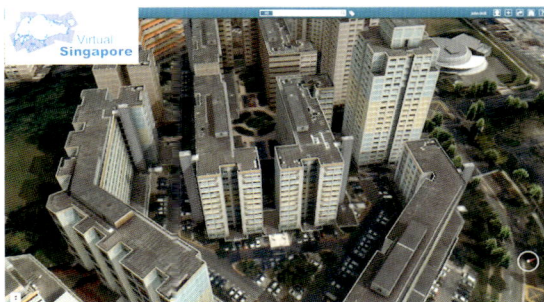

Figure 1.8　Digital City of Virtual Singapore

Figure 1.9　Digital Terrain Model

nents, surface modeling and grid creation of the building on the BIM platform according to specified logical relations, to generate a complete and accurate digital design model.

With the aid of parametric design technology, the digital designer can define the information of factors that affect architectural design as parameters, and incorporate these parameters into a complete design system, analyze and optimize the functions in the design process, and adopt proper algorithms, so as to develop different architectural design proposals and finally select the superior design.

Parametric design method is characterized by strong logic and highly integration. Therefore, compared with general design method, it can develop a model that features clear definition, excellent logic, high accuracy, diverse forms, consistency and highly integration. Parametric technology can help the designer quickly build a high-precision digital building model, establish linkage relations among key model parameters, conduct comparison selection and perfection reconstruction on building proposals, and fully improve the design level and quality of the project.

II. Precise Analysis of Digital Technology

Digital technology not only excels in model generation, but also possesses incomparable advantages over traditional methods in project optimization.

Digital technology can integrate the latest achievements of BIM, IoT and machine learning, build an optimization objective function by virtue of the high-speed operational capability of computer system, replace physical experiments with numerical analysis, continuously modify iterations to improve the system's operation efficiency, and provide technical support for optimization and upgrading of the entire construction industry chain.

The CFD technology can help simulate the air distribution and contaminant dispersion in the building to optimize the layout of ventilation system, establish a city-level large-scale fluency model to simulate the influence of the surrounding wind environment and contaminant dispersion, thus providing design basis for site planning and building layout (Figure 1.10, Figure 1.11).

Developing an effective emergency evacuation plan to scientifically evacuate the crowd

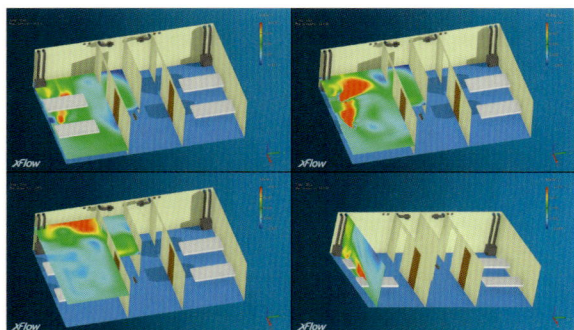

Figure 1.10   Simulation of Contaminant Dispersion in the Building

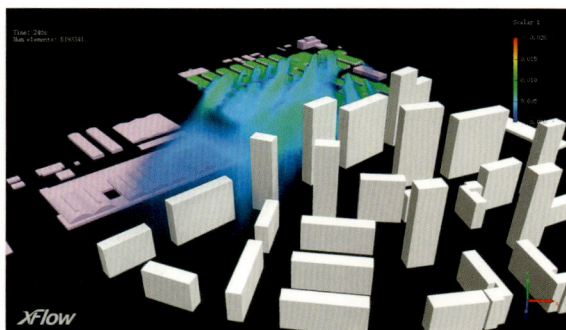

Figure 1.11   Simulation of Contaminant Dispersion at the Site

during rush hours and emergencies has gradually become a complex problem for large-scale public places or important infrastructure with complicated facilities and huge crowds of people. The traditional crowd evacuation plan is mainly formulated based on experts' experience. Digital technology can simulate the crowd evacuation process and display the crowd evacuation scene by establishing the numerical model, optimize spatial layout in the architectural design stage, and formulate proper evacuation plans in the architectural service stage, thus allowing to conduct analysis and research in a more comprehensive, accurate and detailed way. Supported by digital technology, the crowd evacuation simulation possesses many advantages, such as intuitive form, high cos efficiency, low personnel risk and rapid feedback (Figure 1.12) .

Through digital technology, we can calculate and simulate the sun's moving trajectory in the BIM model, calculate the indoor and outdoor sunshine duration and distribution in real time, and optimize the building layout, so as to better meet the requirements of specifications (Figure 1.13) .

Digital technology is widely used to offer favorable technical support for cross-specialty coordination. Supported by IFC, COBie and other standards, it can create data interface between professional softwares, set up the data linkage mechanism of professional models, build a multi-specialty integrated design platform, thus providing more efficient methods and creating better conditions for rapid analysis and structure optimization of building scheme (Figure 1.14) .

III. High-quality Management of Digital Technology

In the engineering design stage, the multi-specialty 3D model and the collaborative design platform can be quickly established to perform design management. Digital technology can quickly check any error, omission, collision and defect in engineering design, and observe and roam in a 3D scene at any time to discover design defects and propose solutions, which will greatly reduce the reworking during construction and avoid the wastage of resources, time and cost.

Digital technology can collect all information of the whole construction process such as

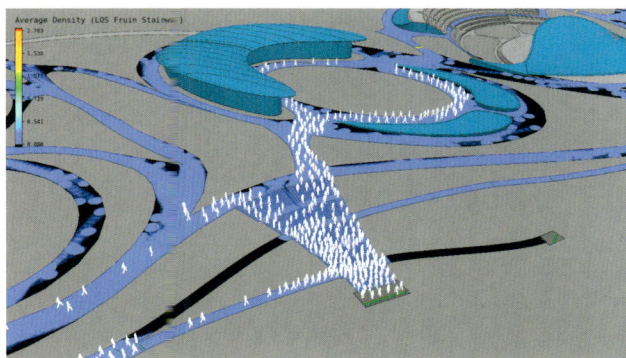

Figure 1.12   Analysis of Evacuation

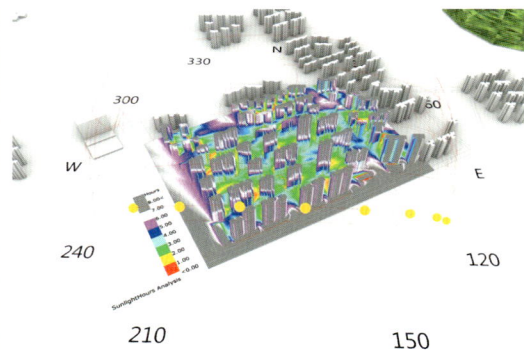

Figure 1.13   Site Sunshine Analysis

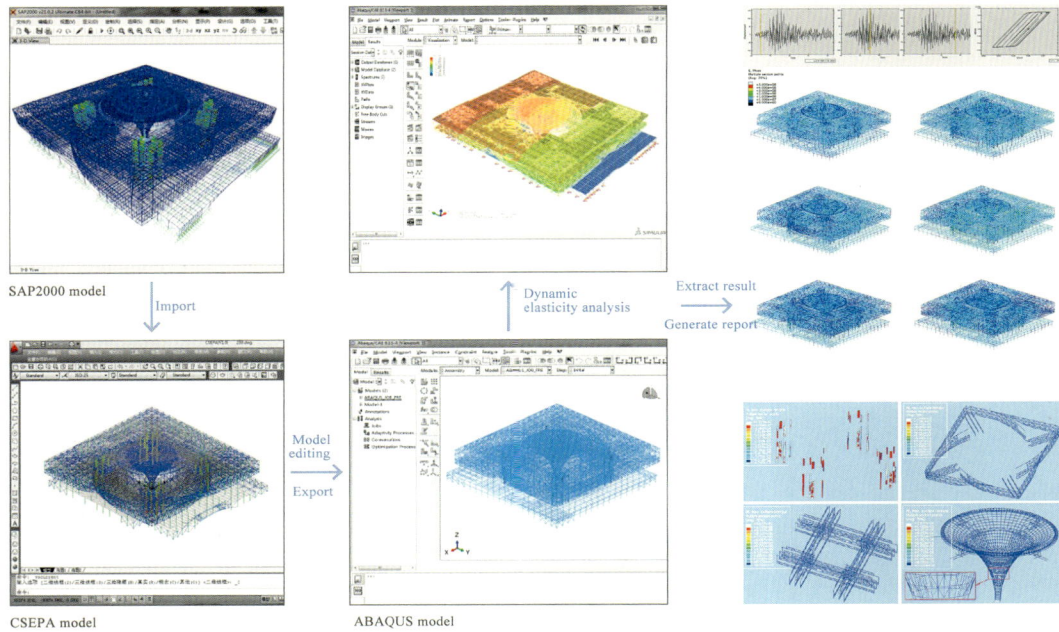

Figure 1.14　Model Conversion

3D models, drawings, contracts, and documents, to form a single data source, and link all the parties (including the owner and the developer, the general contractor, the design institute, the construction contractor, the property management, the building material manufacturers, and equipment suppliers) involved in a project. It also can manage the model documents, data storage, materials and equipment, communication, scheduling, cost control and quality control related to this project in an environment full of unified information exchange and collaborative work, thus achieving efficient coordination and high-quality management of the full lifecycle of buildings (Figure 1.15) .

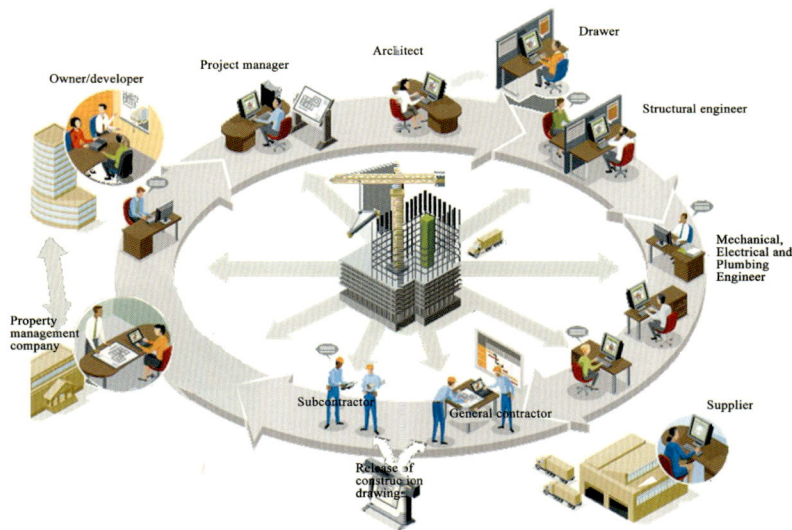

Figure 1.15　High-quality Management of Full Lifecycle of Buildings Based on a Single Data Source

The construction is integrated with the digitization through digital technologies including graphics technology, recognition technology, video surveillance technology, mobile Internet technology, and IoT technology, to collect real-time information at the construction site, automatically complete the progress tracking, materials management, manual scheduling, early warning of key nodes and funds management, and achieve comprehensive analysis and efficient management (Figure 1.16).

1.1.5 Digital Technology Enablement

With the high integration of construction industry and digital technology, the construction industry witnesses some profound changes in its development process, management mode, production mode, supply chain, etc. Digital technology will accelerate the innovation of construction methods, update, transform and upgrade the industry chain of the construction industry, and reshape the business model of the construction industry.

Firstly, digital technology will make the construction industry more productive. The prefabrication ratio of the construction industry has been rising year by year, and the standardization, modularization and prefabrication in the construction industry are increasingly popular. As intelligent manufacturing and automatic delivery are widely spread, the replacement rate of manual work in the whole construction industry will gradually rise. The gradually improved accuracy and efficiency will ensure the construction product quality via the processing mode of production line, and basically eliminate the common quality defects, occupational injury risks, and safety accidents, because a lot of dangerous operations are undertaken by artificial intelligence machines.

The accurately positioned 3D model, directly docked with the automatic production line, can directly provide the material information including product number, texture, quantity, weight and price, and send the data

Figure 1.16 Digital Technology Supports the Project Construction Management

to the production line in many ways to provide assistance.

With the application of new building materials including self-repairing concrete, aerogel and nano-materials, innovative construction methods such as 3D printing and pre-assembly based on 3D building models will be gradually put into use, thereby reducing costs, speeding up construction, and improving quality and safety.

Secondly, digital technology enables the buildings to achieve intelligent perception and continuous optimization. As an outcome of the integration of digital technology and architectural art, intelligent buildings will complete data interchange and information collection through sensors and the internet of things, so as to possess continuous and comprehensive perception. Then, the AI "brain" thinking and decision-making will endow the intelligent buildings with vitality through continuous feedback, self-optimization and iterative update, so that the intelligent buildings can change with the development and change of the organization and people, interact with people, possess perception and make predication at any time, and actively respond to the personalized demands of people.

Finally, digital technology endows the construction industry with a complete digital ecosystem. The upstream and downstream of the construction industry chain are composed of specialized companies, such as prefabricated parts manufacturers, material distribution centers, electromechanical equipment companies, and professional suppliers. By building the industrial Internet of the construction industry and linking the companies in the construction industry chain, the general contractor and the owner jointly perform all responsibilities in investment, design, construction, operation, and maintenance throughout the full lifecycle of buildings, make use of the industrial Internet to find suitable suppliers. Meanwhile, the downstream enterprises of building industry chain can realize interconnection and resource sharing through the building lifecycle management platform, and jointly create a complete business and management system. In this way, the intelligent operation management system works efficiently. The production efficiency and benefit are significantly improved, and energy conservation and low-carbon environmental protection of buildings reach an unprecedented level.

1.2 Modularization, Standardization and Prefabrication : Rapid Construction of Leishenshan Hospital

On the afternoon of January 25, Wuhan government decided to build Leishenshan Hospital on the south of the Yangtze River within about 10 days, which is larger than Huoshenshan Hospital standing on the north of the river. The gross building area of Leishenshan Hospital has been adjusted up for three times, and the final scale is determined as : 80 000 m² gross building area and 1 500 beds. The hospital can accommodate about 2 300 medical staff. At 8 : 00 p.m. February 8, Wuhan Leishenshan Hospital was completed and handed over, and received the first batch of COVID-19 patients.

The overall scale of Leishenshan Hospital doubles that of Huoshenshan Hospital, but its construction schedule is tighter than that of Huoshenshan Hospital. In addition to the split-second round-the-clock selfless work of participants, and scientific and reasonable design & construction organization, there are three important guarantee factors for the rapid construction of Leishenshan Hospital, that is modular design, standardized production, and prefabricated construction, as shown in Figure 1.17.

1.2.1 Modular Design

The concept of modularization has been applied in the design of Leishenshan Hospital and can be summarized into three major levels: macro planning framework, meso-level functional layout, and micro room design.

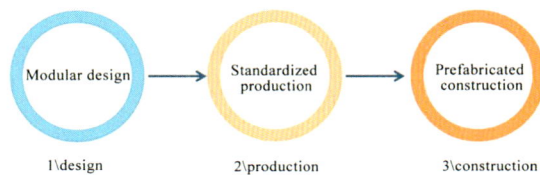

Figure 1.17　Hook-ups: Three Stages for Construction of Leishenshan Hospital

I. Modular Planning Framework

Leishenshan Hospital is divided into three major modules: isolation medical area, living area for medical staff, and support area, of which the isolation medical area provides the main function, as shown in Figure 1.18.

According to the terrains, the isolation area in the fishbone layout is divided into the north area and the south area. Each area takes the medical staff passage as its central axis, and ward units and test and imaging units are connected to the central axis as functional modules. The north area and the south area each have 15 ward modules, which are set every 12 m. Other functional modules are as follows: the 30-bed ICU modules (each for the north area and the south area), clinical lab module, operating room & CT room module, pharmacy and drug store module, and centralized sanitary passage modules (each for the north area and the south area). In total, the north area has 20 functional modules, while the south area has 17, as shown in Figure 1.19. Among them, ward modules are the main part of the isolation area, and each ward module is equipped with 50 beds. All ward modules have

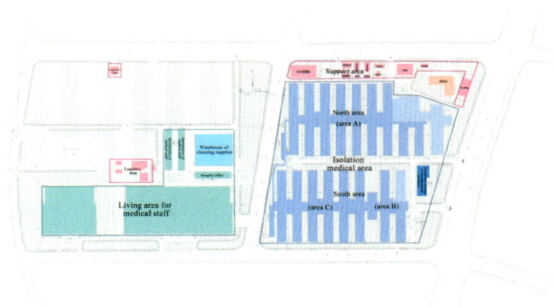

Figure 1.18　Overall Functional Areas of the Project

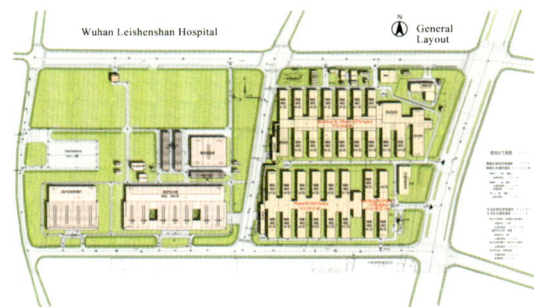

Figure 1.19　General Plan of Project: Fishbone Layout of Isolation Medical Area

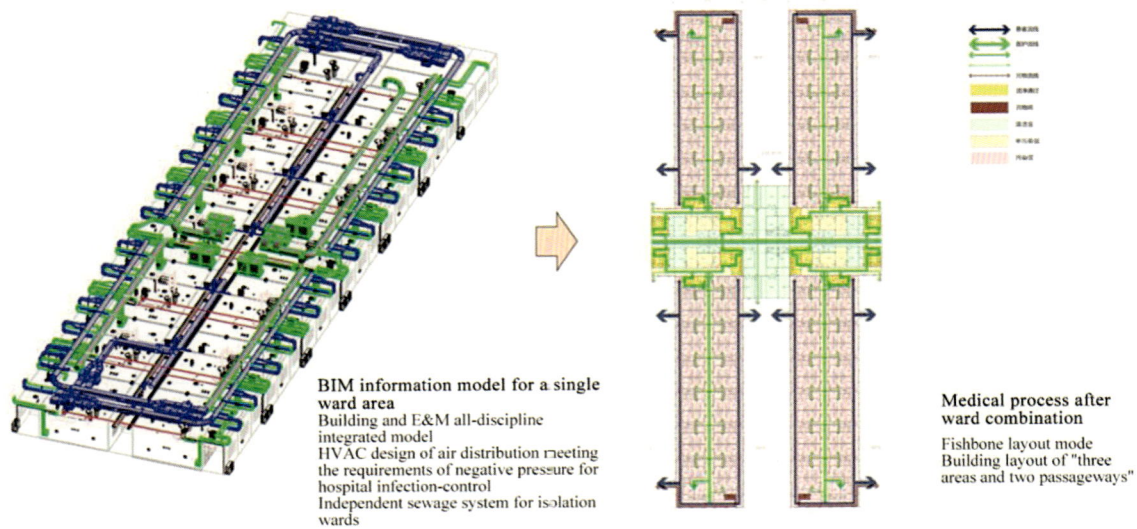

BIM information model for a single
ward area
Building and E&M all-discipline
integrated model
HVAC design of air distribution meeting
the requirements of negative pressure for
hospital infection-control
Independent sewage system for isolation
wards

Medical process after
ward combination
Fishbone layout mode
Building layout of "three
areas and two passageways"

Figure 1.20   Ward and Treatment Area Module and Its Medical Process

universal and consistent design, and follow the principles of air distribution "three areas and dual passages", which is critical for infectious disease wards, as shown in Figure 1.20.

Each module has a clear boundary and independent functions, and is connected with each other through the central axis. This facilitates simultaneous design and separate construction, and saves the design and construction time.

II. Modular Functional Layout

At the meso level, the medical function can be quickly realized through the modular functional layout. For the purpose of reducing the types of functional modules, minimizing construction differences, and improving construction efficiency, 95% medical area of Leishenshan Hospital is mainly composed of three basic functional modules, namely :

(1) Basic ward module (i.e. functional module for negative pressure wards in the contaminated area), consisting of two wards and a shared front buffer room.

(2) Working area for medical staff (i.e. functional module for semi-contaminated area), composed of sanitary passages for male and female medical staff, nurse station, dispensing room, doctor's office, and transfer room.

(3) Auxiliary medical area (i.e. functional module for hygienic area), composed of duty room, restroom for medical staff, drug store, instrument store, consumables store, and distribution room.

The three basic functional modules are assembled into 30 ward areas as per unified internal standards. The three basic functional modules also map to three important basic areas in treatment of infectious diseases : contaminated area, semi-contaminated area, and hygienic area. Through the fine design of the three basic functional modules, the medical processes, construction methods, and E&M pipelines of each functional module can not only meet the high-standard process requirements of emergency hospitals, but also facilitate simple and quick construction and E&M installation. Only when the basic modules at the bottom are properly completed, can the large-scale Leishenshan Hospital maintain a high overall quality in the rapid construction process, and meet the high-standard medical process requirements of emergency hospitals, as shown in Figure 1.21.

III. Modular Room Design

The design is broken down to each room. For convenience of standardized production, the whole isolation ward and treatment area (including the pharmacies, drug stores, and centralized sanitary passages) is divided into two frame modules A and B for container house via modular partition, in which module A has a size of 3 m × 6 m × 2.9 m ($W \times L \times H$) and a total number of 1 918 (970 in the north area and 948 in the south area) ; module B has a size of 2 m × 6 m × 2.9 m ($W \times L \times H$) and a

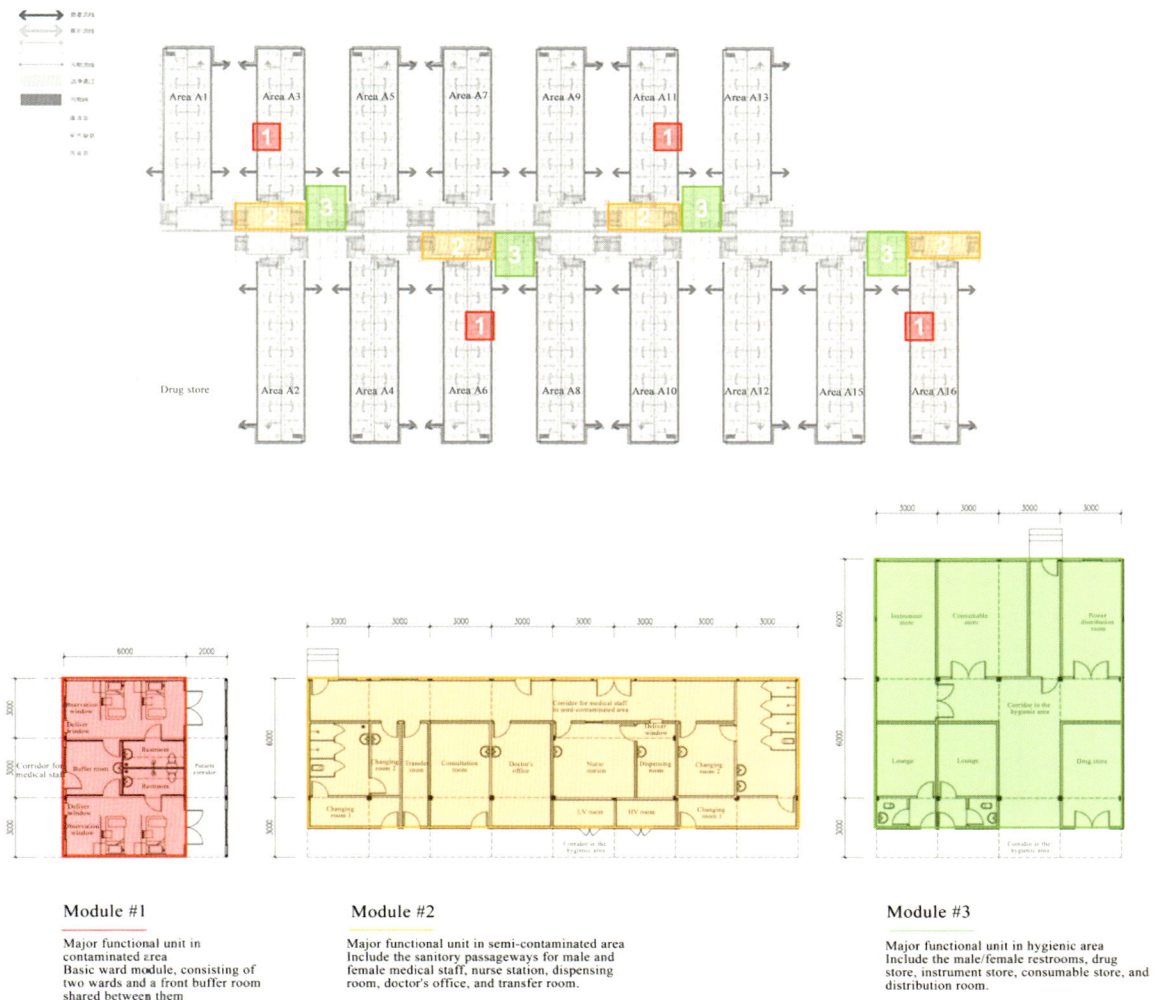

Module #1

Major functional unit in contaminated area
Basic ward module, consisting of two wards and a front buffer room shared between them

Module #2

Major functional unit in semi-contaminated area
Include the sanitary passageways for male and female medical staff, nurse station, dispensing room, doctor's office, and transfer room.

Module #3

Major functional unit in hygienic area
Include the male/female restrooms, drug store, instrument store, consumable store, and distribution room.

Figure 1.21　Three Basic Functional Units

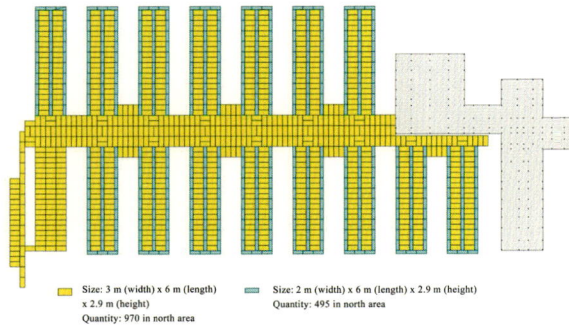

Size: 3 m (width) x 6 m (length) x 2.9 m (height) Quantity: 970 in north area

Size: 2 m (width) x 6 m (length) x 2.9 m (height) Quantity: 495 in north area

Figure 1.22 Dimensions and Layout for Two Types of Container Housing Modules in the North Area

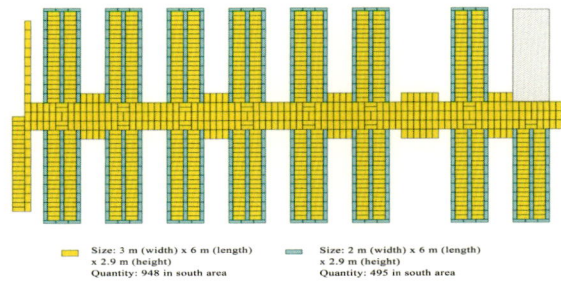

Size: 3 m (width) x 6 m (length) x 2.9 m (height) Quantity: 948 in south area

Size: 2 m (width) x 6 m (length) x 2.9 m (height) Quantity: 495 in south area

Figure 1.23 Dimensions and Layout for Two Types of Container House Modules in the South Area

total number of 990 (495 in the north area and 495 in the south area), as shown in Figures 1.22 and 1.23. All modules are prefabricated in factory, transported to the roads around the site for assembly, and hoisted in place on site. Small pipelines inside the ward module are reserved and embedded simultaneously. Some special functional rooms (such as pharmacies and drug stores) are built in container units to form spacious space through different combinations in horizontal and vertical directions.

Furthermore, modular design focuses on major functional rooms, especially the rooms closely related to the use of medical functions and processes in emergency hospitals, which mainly includes： ① basic ward modules, as shown in

Figure 1.24 ; ② distribution rooms, small built-in restrooms, and other equipment rooms, as shown in Figure 1.25 ; ③ sanitary passages connecting different areas, and other critical hospital infection-control rooms, as shown in Figure 1.26.

1.2.2 Standardized Production

On the basis of modular design, the standardized production of components has been carried out simultaneously in many factories across the country, amid the field processing of terrace and embedded pipe network. Except for foundation treatment, main pipe network embedment, and concrete exterior wall of CT room, all works on the site are subject to dry operation. Over 95% building materials of

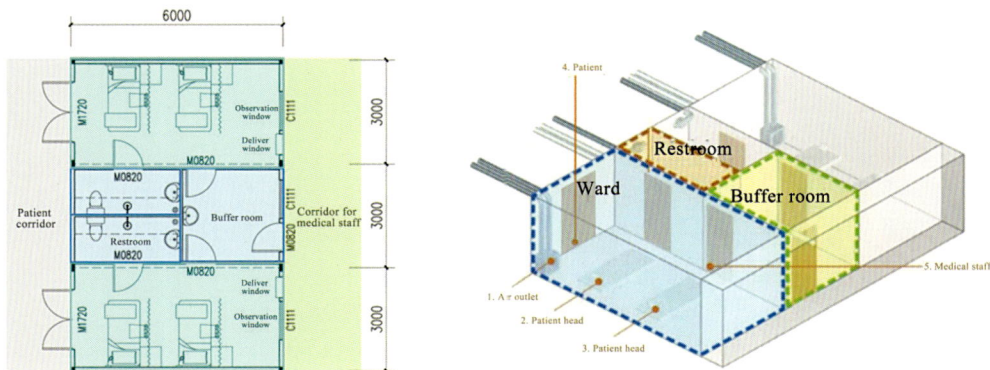

Figure 1.24 Basic Ward Modules—Satisfying Doctor-Patient Shunt, Buffer Passage, and Airflow Distribution

Design and Construction Technology of Digital Emergency Hospital

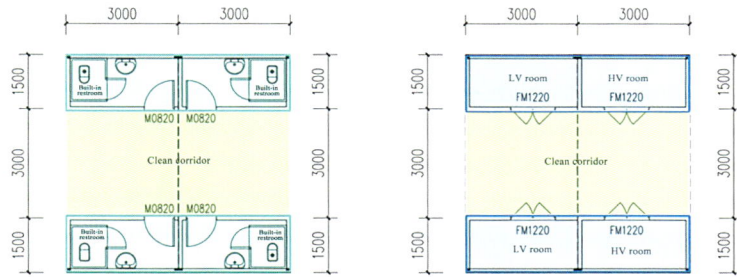

Figure 1.25 Distribution Rooms, Small Built-in Restrooms, and Other Equipment Rooms

Left: small restrooms for medical staff, totally 7 in north and south areas; Right: distribution rooms, totally 16 in north and south areas

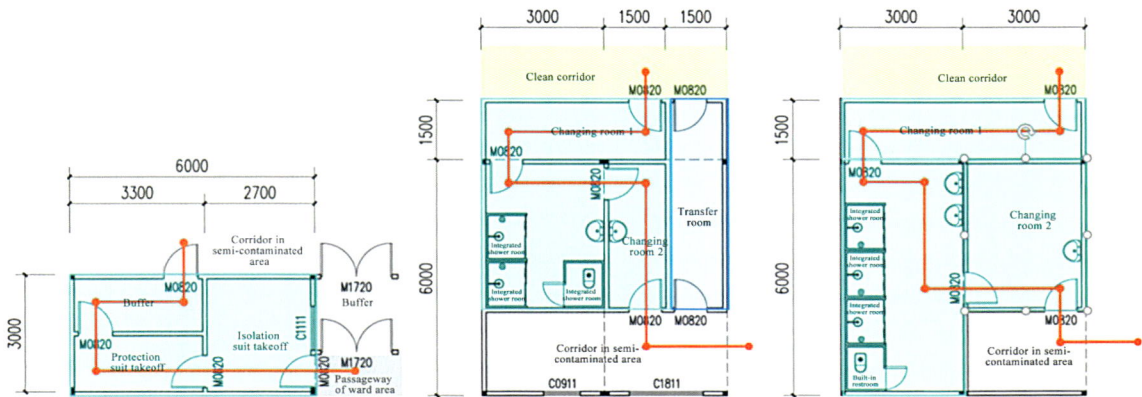

Figure 1.26 Sanitary Passages Connecting Different Areas, and Other Critical Hospital Infection-Control Rooms

Left: takeoff buffer room, totally 30 in north and south areas; Middle: sanitary passage for male staff, totally 32 in north and south areas; Right: sanitary passage for female staff, totally 32 in north and south areas

Leishenshan Hospital are prefabricated in factories, such as support of prefabricated houses in the medical area, modular container house of steel structure (Figure 1.27), doors and windows, steel structure components (Figure 1.28), finished restrooms and shower rooms, inner wall partition in test and imaging area, sterile wallboard for operating rooms and clinical labs, and all bathroom products and E&M facilities.

Figure 1.27 A Manufacturer from Jiangsu Province Providing Aid by Building the Container-Type Prefabricated House for Leishenshan Hospital

Figure 1.28　A Steel Structure Manufacturer from Anhui Province Only Taking 40 Hours to Load and Transport 2000 Sets of Standardized Steel Structural Components to the Site of Leishenshan Hospital

Based on unified technical specifications, standardized and industrialized production, and satisfactory factory environment and processing precision, operators can control the quality of components, minimize defects, and ensure the overall construction quality during rapid field assembly. Reducing wet operations on site allows simultaneous installation of multiple services. Moreover, standard prefabricated components can be recycled, disassembled and reassembled, thus saving energy and protecting the environment.

1.2.3　Prefabricated Construction

The major challenges in the construction of Leishenshan Hospital are its tight schedule and heavy task, which requires completing a 2 or 3 years construction task within about 10 days. Prefabricated construction is an inevitable choice for project construction. This project incorporates three different prefabricated construction methods according to different building functions and floor height requirements for the inpatient department and test and imaging department in the isolation area, and living area for medical staff.

I. Prefabricated Construction Method for Inpatient Department

The inpatient department in the isolation area has the most functional rooms in the whole project, accounting for about 70% of the total. All these rooms are built into mature modular container houses of steel structure (container-type prefabricated house) as the basic construction units. Compared with traditional color steel prefabricated houses, container-type prefabricated houses perform better on fire prevention, earthquake resistance, sound insulation, sealing, waterproof, and heat preservation. All modular components are subject to standardized production in factories, and building components are quickly installed on site with standardized connectors to complete the overall building, as shown in Figure 1.29.

II. Prefabricated Construction Method for the Test and Imaging Department

The test and imaging department in the isolation area mainly includes ICU, CT room, operating room, and checkout room. The test and imaging building, as a comprehensive building for examination, diagnosis, and treatment, serves as the "brain" of Leishenshan

Figure 1.29　Field Assembly of Steel Structure Box-type Modular House (Container-type Prefabricated House)

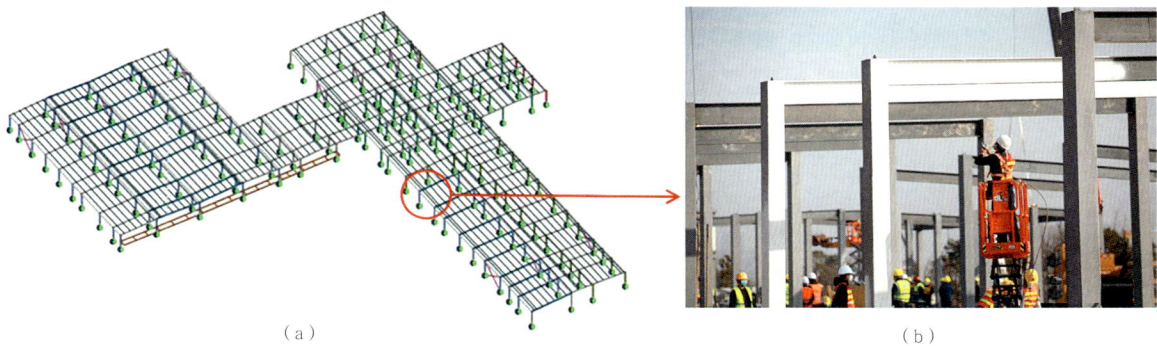

(a)　　　　　　　　　　　　　　　　　　　　　　(b)

Figure 1.30　Steel-Frame Structure System for Test and Imaging Department
（a）Structure Diagram of Test and Imaging Area（Midas Analysis Model）；（b）Field Assembly of Steel Structure Prefabricated Components for Test and Imaging Area

Hospital, which holds complicated functional rooms with large span and floor height, and non-uniform plane column grid. The structural span reaches over 7 m, and local span 18 m, and floor height over 4.2 m. The steel-framed structure system is used because it is difficult for container house and steel-framed house with sandwich panel systems to meet the requirements. Steel structure design and detailed node design are carried out simultaneously, and steel structural components are prefabricated in factories and transported to the site for assembly, as shown in Figure 1.30.

III. Prefabricated Construction Method for

Living Area for Medical Staff

According to the plane and space requirements of the building, the living area for medical staff is of a two-storey light-steel prefabricated house system, which is different from the ward area. The plane layout of building takes the wall board width—1 820 mm as the module, and the plane size of a standard room is 3 640 mm × 5 460 mm. The main structure of the light-steel prefabricated house is a light steel frame with cross stay cables to ensure structural rigidity and stability. With mature technology, standardized modular design and convenient assembly and disassembly, prefab-

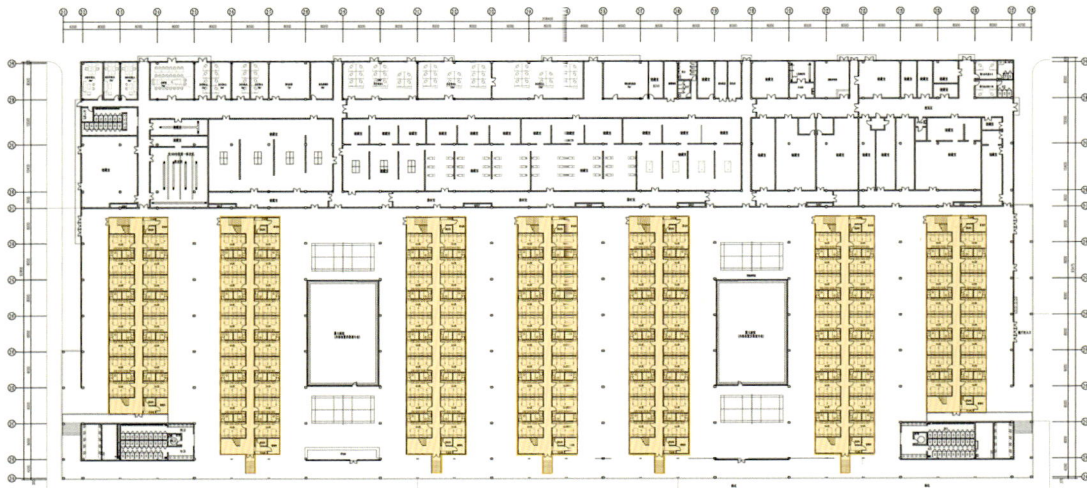

(a)

(b)

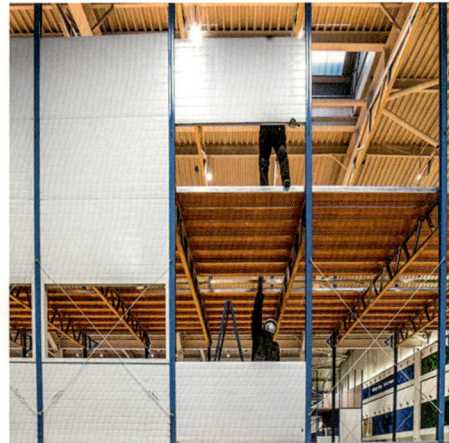

(c)

Figure 1.31　Light-Steel Prefabricated House Area and Construction System for Living Area for Medical Staff
（a）Building Plan；（b）Light-weight Steel Temperary House Area；（c）Construction System

ricated houses show obvious advantages like flexible plane layout when serving as indoor temporary buildings, and can satisfying functional requirements of the building, as shown in Figure 1.31.

1.3　Structural Design of Leishenshan Hospital

1.3.1　Design Principles

As a result of the emerging circumstances under COVID-19, the key problem comes to how to construct the hospital within an extremely short time. In this project, the prefabrication construction method is used to effectively shorten the building period. While the light modular system can fulfill the requirements on the functionality and spaces of the temporary hospital, it is also less demanding on the site conditions as compared to the conventional construction method. The structural design of Leishenshan Hospital, which is a prefabricated building, follows the principle of standardization, modularization and integration, while utilizing the existing products in industry as much as possible. In general, the higher the integration level of the prefabricated modules, the less demanding will be on the field installation, which means faster construction. This ultimately leads to a better guarantee on the quality of the finished work.

1.3.2 Superstructure

Different structural types are selected for the isolation area and living area for the medical staff, according to their various functions and spatial characteristics.

I. Isolation Area

The isolation area can be categorized into two typical units : the ward care unit and the test and imaging unit.

1. Ward Care Unit

Ward care units and medical office units are featured by their unified size and dimensions, which corresponds to the characteristics of standardization and modularization. Therefore, we adopted the light steel structure with prefabricated modular units (container houses). Each of these assembled units are composed of modular floors, roofs and wallboards. Also, the mechanical pipelines, doors, windows and decorative parts can be incorporated along into the module unit. Moreover, in addition to the building performance specifications, these modular units should also fulfill requirements in the transportation, lifting and installation process.

As for the building materials, cold-formed thin-walled sections are largely used for the steel frames as well as the color steel laminboards for the walls. These modular units can be transformed and spliced freely as required, by taking each container as the basic unit. They may be used alone, or to be combined together to form a much spatial area. This can be achieved by removing the partition walls between and connecting the units through different alignments, both horizontally and vertically.

2. Test and Imaging Unit

The ICU, test and imaging area has floor with high elevation. Since they typically have non-uniform column grid alignments with larger span, the container house or prefabrication one can not meet the design requirements. As a result, we adopted the steel frame structural system. The color steel laminboards are still used for the enclosure and partition walls, in order to facilitate the installation, whereas the column grids are sized depending on the available dimensions of boards. As for connection to the roof, the standing seam interlock was used on the insulation roofing system.

II. Living Area for Medical Staff

The living area for medical staff consists of a building with two-storey structure. According to the architectural plan and space requirements, we choose the light steel prefabricated building system. The structural layout was configured based on available dimensions of the wallboard.

In the structural design of this area, the light steel prefabricated houses are preferred over container modules, for their higher flexibility in fulfilling the demands from various functions.

1.3.3 Foundation Design

I. Geotechnical Conditions

The building on site mainly uses silty clay as the bearing stratum, which is featured by its high bearing capacity and relatively large compression modulus.

II. Foundation of Isolation Area

By taking advantage of the original hardened pavement on the parking lot as much as possible, we erect structural steel piers or raft foundation on the ground, according to specific layout of the superstructure. On one hand, this solves the problem of foundation cushion. On the other hand, it facilitates the installation of HDPE impermeable layer (high-density polyethylene film) across the site.

Since the ward care unit containers are generally light-weighted, they can be directly placed on the hardened pavement. But in order to install the drainage pipe in a fast way, we use short steel piers to lift the container houses at a certain distance above the ground.

For the Test and Imaging units which are supported by steel frames, large reaction forces can be found at the column bottoms. For the hardened ground in-situ may fail locally under the concentrated loads, a 300 mm thick reinforced concrete layer is added on top of the pavement, creating a superimposed flat-slab-type mat foundation.

III. Foundation of the Living Area for Medical Staff

The dormitory for medical staff is located inside the 10 000-People's-Dining-Hall and is placed on the hardened pavement. To fa-

cilitate the installation of drainage pipes, we set up an empty space beneath the dormitory building with a height of 1.7 m. The finished products of type-321 Bailey beams are used in this purpose, forming an empty space for the prefabricated building structure.

1.3.4 Coordination in Construction

I. Construction Process

Leishenshan Hospital project was commenced on January 25, 2020 and handed over on February 6, 2020, with construction period of only about 10 days. The construction process can be mainly illustrated as: ① land grading; ② trench excavation for pipes and backfilling; ③ soil hardening and flooring; ④ HDPE membrane installation; ⑤ construction of the strip footing, structural steel piers and the raft foundation; ⑥ installation of container houses and steel frames; ⑦ installation of enclosure and partition wallboards, plus roofing system; ⑧ pipeline installation and decoration; ⑨ set-up and commissioning of medical equipment and furniture.

II. Key Issues on Coordination during Construction

Structure designers are required to integrate the resources available to the contractor in construction, and use building materials with dimensions provided. One should only propose the structural design with proper implementation on site, after the designer have a good master of the field and work conditions. When it becomes difficult in implementing the original scheme due to unexpected changes, designers should help adjust the design in time and cooperate with the construction contractor. By recognizing some building components might not be finished on time, or the construction procedure may not be followed exactly on site, structural engineers should deal with these problems with flexibility and take effective measures to ensure the safety of the temporary structure, while not delaying the construction as a whole.

1.3.5　Summary

In conclusion, Leishenshan Hospital is a heavy task with high requirements. Therefore, it is designed and constructed simultaneously. Owing to the high efficiency in design and effective construction management and coordination, and thanks to the well-developed industry of light steel prefabrication structure, Leishenshan Hospital was completed successfully in an extremely short period. Based on the experience in design and construction coordination, we propose the following suggestions for future reference.

（1）In the preliminary stage of design, it is advised to sufficiently communicate with the construction contractor, in terms of duration, processing and transportation, labor and equipment, as well as material supply and construction methods, so as to minimize the changes on site and modification on design.

（2）In future, designers can prioritize the prefabrication building structure when they design

temporary hospitals. It is required to follow the principles of standardization, modularization and integration, while making good use

of the existing construction industrial product system.

（3）Simplified design process, with simultaneous coordination with the construction on site. Adopt structure types and foundation with stronger adaptability.

（4）Structural designers should consider the space requirement and self-weights of the MEP.

（5）The structural joints should be simple in configuration and reliable, and easy to be connected with the other structural components.

（6）More effective waterproofing at the splice joints of the modular components.

（7）During the fast construction, safety and risks should be top concern and be responded in time with proper measures.

1.4　Summary and Thinking on Water Supply and Drainage Design of Leishenshan Hospital

1.4.1　Project Profile and Design Portfolio

Leishenshan Hospital sits to the north of Qiangjun Road, Junyun Village, Jiangxia District, Wuhan, with a construction area of 220 000 m^2 and a gross building area of 80 000 m^2. The dining hall of Junyun Village was originally planned on the west, and a central parking lot on the east（Figure 1.32）.

The Leishenshan Hospital is designed as per the standards for emergency hospitals, to receive the confirmed COVID-19 patients. According to the land layout, the east and west areas are respectively planned as the isolation

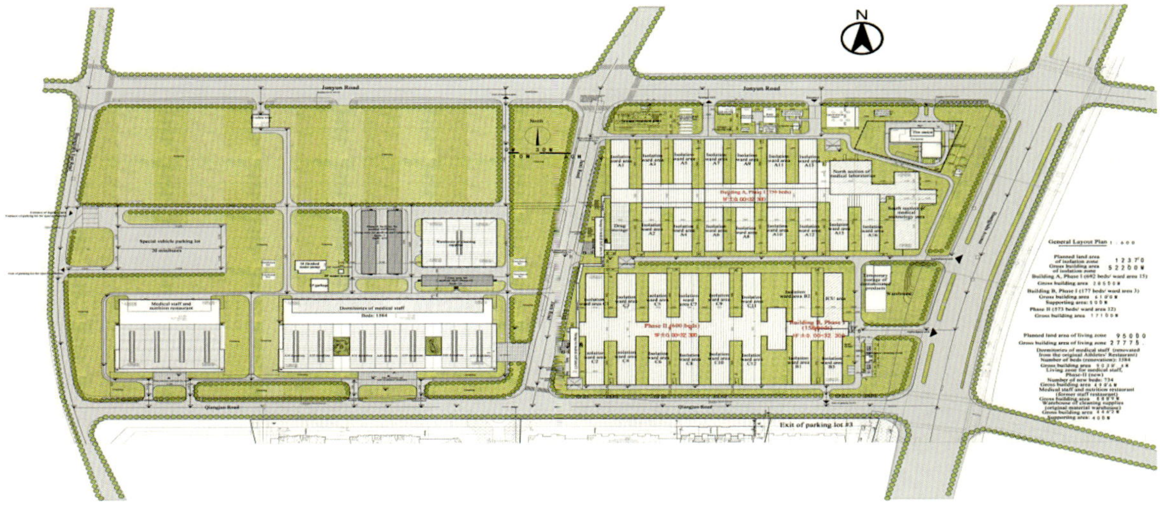

Figure 1.32　General Outdoor Layout of Leishenshan Hospital

area and living area for medical staff, and are provided with O&M rooms. The hospital can hold a total of 1500 beds and accommodate about 2318 medical workers.

The project has the maximum domestic water consumption of 1 140 m^3/d and the maximum water discharge of 1 083 m^3/d.

This design includes: outdoor water supply & drainage design, outdoor firefighting design, indoor water supply & drainage design, indoor firefighting design, sewage treatment and disinfection.

1.4.2　Determination of Outdoor Drainage Scheme

Leishenshan Hospital is built on the large parking lot and dining hall of Junyun Village originally planned for the Military World Games, and the original site already has 300 mm thick concrete rigid floor, so the hospital can be directly built on the site without site leveling.

I. Excavation of Pipe Trench

For laying of outdoor drainage pipes and minimizing the excavated volume of the construction contractor, all outgoing drainage pipes are laid in the 400 mm cushion back-filled on the hardened site. Outdoor rainwater, sewage, and wastewater main pipes between isolation ward units are laid in 2.5 m wide excavated trenches, and all pipes have consistent slope and bottom elevation for convenience of excavation and construction. The rainwater, sewage, and wastewater main pipes on other arterial roads are also laid side by side in excavated trenches to reduce the excavation surface. In order to reduce the work of pipe trench excavation during construction, two units share one outdoor drain header, reducing by half the work of outdoor pipe trench excavation between isolation ward units, and greatly shortening the construction time of outdoor pipe network.

II. Sewerage System

Outdoor sewage from ward and treatment area and other areas, and outdoor rainwater are separately discharged through independent drainage pipes.

III. Selection of Pipe Materials

PE solid wall pipes are used as outdoor drainage pipes for hot welding. The pipe foundation is made of 100 mm thick C15 concrete cushion, with 150 mm thick fine sand laid above.

For pipe diameter ≤ 600 mm, the finished plastic inspection well is adopted; for pipe diameter > 600 mm, the prefabricated reinforced concrete inspection well is adopted. Sealed manhole cover is used.

1.4.3 Domestic Water Supply and Drinking Water System

I. Domestic Water Supply System

Break tank is used for water supply. The domestic water tank has a capacity of 300 m^3 and stores 25% of the maximum daily water consumption. To improve its reliability for water supply, one water supply pipe is routed from different municipal roads to the inlet of the domestic water tank respectively.

Domestic water is supplied by a VF pressurized pump unit under the control of one-to-one frequency converter, and the outlet pipe of the water pump in the pump room is equipped with a UV disinfector co-working with the anti-fouling sterilizer.

Water supply networks are set up for the ward and treatment area and other areas respectively, and a backflow preventer is set up at the front end of the water supply header in the ward and treatment area. Each ward and treatment area is provided with control valves for water supply, and such valves are set in a place easy to access within the hygienic area. The outdoor water supply network is of loop shape.

II. Supply of Drinking Water

Boiler rooms are arranged in a scattered layout, and two 12 kW boilers with filtering devices are installed in each boiler room, which can provide boiled water and normal temperature water.

1.4.4 Indoor Hot Water System

The project has a large demand on hot water. According to the project characteristics and manufacturer's donation, domestic hot water is provided via different methods.

I. Hot Water Supply to the Ward and Treatment Area

Each ward accommodates two persons at most and is equipped with the electric water heater to supply hot water, which not only meets the patients' need, but also avoids cross infection.

II. Hot Water Supply to the Medical Area

The medical area has centralized bathrooms and a large number of doctor showers. In view of the natural gas supply available in the medical area and the demand and characteristics of hot water consumption by doctors, it is appropriate to adopt the commercial water heaters for central supply of hot water to showers in the medical area.

1.4.5 Indoor Sewerage System

I. Sewerage System

Sewage from the ward and treatment area and other areas is separately discharged through independent drainage pipes.

II. Safeguard Measures for Water Seal of Drainage Pipes

（1）The main drainage pipes that connect the horizontal branches of every toilet is comparatively long, so a loop vent is set at each branch's end, to maintain the drainage capacity of the pipe and protect the water seal.

（2）Floor drains are provided in the preparation room, filth cleaning room, restroom, bathroom, and air conditioning room, but not provided in the nurse room, treatment room, consulting room, clinical lab, and doctor's office. The floor drain is of non-water seal type with filter screen and trap, and the trap has a water seal of 50 mm. Floor drains in the operating room and emergency room is of openable sealed type.

（3）Drainage from the wash basin replenishes the water seal of the floor drain.

III. Other Safeguard Measures

（1）Several loop vents of each isolation medical unit merge, penetrate the side wall, and ascend to the roof, and an ultraviolet air sterilizer is installed at the end of the exhaust pipe. The outlet of the vent pipe is kept away from the fresh air room on the roof.

（2）Condensate from air conditioners is indirectly discharged to floor drains and enters the sewerage system of the hospital.

（3）All wash basins and sewage tanks are not provided with plugs.

1.4.6 Sewage Treatment and Disinfection

The sewage treatment plant is located to the north of the land parcel and mainly treats all sewage and wastewater from the isolation area. It is designed according to the parallel operation of two 40 m^3/h treatment units, with a total treatment capacity of 80 m^3/h.

I. Sewage Treatment Technology

Leishenshan Hospital is a specialized emergency hospital, and its sewage is treated by the technology of "contact disinfecting tank + septic tank + lift pump station（with crushing grille）+ regulating tank + MBBR biochemical pool + coagulative precipitation tank + baffling disinfecting tank", which meets the *Discharge Standard of Water Pollutants for Medical Organization*.

II. Disinfection Technologies

The project uses chlorine dioxide for disinfection. The integrated chlorine dioxide equipment is adopted, where sodium chlorate reacts with hydrochloric acid to prepare the chlorine dioxide disinfectant.

The sewage treatment technology provides intensified disinfection twice, i. e. the primary disinfection at the inlet of the contact disinfecting tank, and the secondary disinfection at the inlet of the baffling disinfecting tank. Contact disinfecting tank shall be of closed type, with the water retention time of over 1.5 h and unit chlorine dosage of 40 mg/L（available chlorine）. The baffling disinfecting tank has a water retention time of over 1.5 h and unit chlorine dosage of 25 mg/L（available chlorine）.

1.4.7　Design of the Fire System

I. Safety Guarantee of Fire Water Source

Safety guarantee of outdoor fire water source and perfect outdoor fire hydrant pipe network are prerequisites for the design of fire water system in the project. This is a multi-story building, and in case of fire, firefighters can take water from outdoor fire hydrants for fire extinguishing.

The pressure of municipal water supply network around the project is 0.35 MPa, and a DN200 municipal water supply pipe is routed from Qiangjun Road and Junyun Road respectively to the outdoor firefighting pipe network.

II. Layout of Outdoor Firefighting Pipe Network and Outdoor Fire Hydrants

The outdoor firefighting pipe network is arranged in a loop shape, with outdoor aboveground fire hydrants. These hydrants provide one DN150 mm hydrant mouth and two DN65 mm hydrant mouths. The distance between outdoor hydrants should be with-in 120 m and the outdoor hydrant should be more than 2 m away from roadside (Figure 1.33) .

III. Indoor Firefighting Facilities

The converted buildings make full use of the existing indoor firefighting facilities of the original buildings for the Military World Games and strengthen the layout of fire extinguishers within the buildings. Considering that the construction schedule is tight, and Leishenshan Hospital is for temporary use and it adjoins Jiangxia Fire Station in the north, no indoor fire hydrant system and automatic sprinkling system are provided in new buildings, but the layout of indoor fire extinguishers is optimized. In the emergency room of the hospital, the suspended superfine dry powder automatic fire extinguishing device is provided to strengthen the protection.

1.4.8　Design Thinking

The construction of Leishenshan Hos-

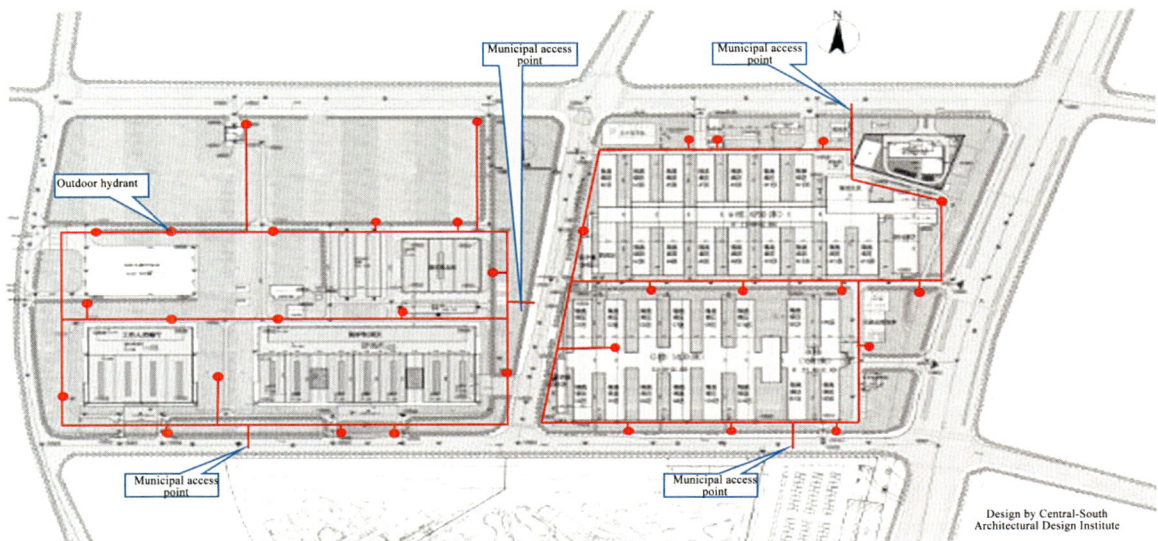

Figure 1.33　Layout of Outdoor Fire Hydrant Pipe Network (Wuhan Leishenshan Hospital)

pital carried out smoothly, and it took about 10 days from design to put-into-use, with construction and design conducted at the same time. The water supply & drainage design team completed the design task perfectly with a high sense of responsibility and mission, and high level of expertise.

Since the design and construction of Leishenshan Hospital is extremely urgent, the indoor and outdoor water supply & drainage scheme should meet the usage demand of doctors and patients, and the requirements for convenient and fast construction in accordance with the construction standards for emergency hospitals. We considered market procurement, product inventory, and manufacturer donation for selection of all materials.

I. Analysis on Key Points and Difficulties

（1）The construction of outdoor drainage pipe network at the early stage directly affects the construction schedule. Therefore, the outdoor drainage scheme must meet the outdoor drainage requirements of emergency hospital, and catch up with the construction schedule of the hospital, by minimizing the site excavation and backfill earthwork, and shortening the construction time of outdoor pipe network.

（2）Since the sewage treatment plant provides many disinfection methods, we compared liquid chlorine, chlorine dioxide, and chloros regarding the selection of disinfectants. After comprehensive comparison, we determined to use chlorine dioxide for disinfection in this design.

（3）The project is a temporary emergency hospital, a densely populated venue. How to grasp the key points of the design standards for fire water system in the hospital poses a great challenge to the fire water system design of the water supply & drainage discipline. Under the premise that the fire water source and outdoor fire pipe network is guaranteed, in order to catch up the construction schedule, the indoor firefighting design, not subject to the limitations of national fire code, makes full use of the existing firefighting facilities in the buildings originally planned for the Military World Games, renovates the existing firefighting facilities, optimizes the layout of fire extinguishers, and enhances fire management measures, to guarantee fire safety.

II. Return Visit to Project at Later Stage

Upon substantial alleviation of the epidemic, CSADI has organized a return visit to Leishenshan Hospital. Practice has proven that, the water supply & drainage design of Leishenshan Hospital is safe and effective, and fully meets the demand of the emergency hospital. However, the following problems need to be further resolved in future design :

（1）The sewage tank and wash basin with a bigger size shall be purchased to prevent sewage splashing.

（2）Valves shall be installed at the positions to make it convenient for overhaul and maintenance.

（3）Hot water shall be available in ICU to meet patients' needs.

1.5 Electrical Design

1.5.1 Power Supply and Distribution System

The temporary emergency hospital should apply two independent mains power supply, with 100% standby power supply . The hospital should be equipped with emergency diesel generator sets. Generator sets shall automatically start and output within 15 seconds in case of mains supply failure.

In addition to the two mains supply, the air-purification operating area, test and imaging area, emergency treatment area, ICU area, negative pressure isolation ward and treatment area, medical laboratory, medical gas equipment, sewage treatment equipment and fire-fighting electrical equipment shall be supported by emergency diesel generator set as a standby power supply.

In order to speed up construction, the outdoor box-type substation and outdoor box-type silent diesel generator set should be adopted and installed in a centralized manner. The generator sets should have its own daily fuel tank and oil supply port.

1.5.2 Low-voltage Distribution System

Except non-critical load such as the heating and air conditioning, other power lines of individual buildings should be of double-circuit automatic switching power supply. Operating rooms, emergency rooms and ICUs should be provided with UPS and medical IT system. UPS should also be provided for laboratory equipment and other medical equipment with no power failure allowed.

The distribution room and electrical shaft should be arranged at the hygienic area. The distribution box should be provided, to the extent possible, at the hygienic area, rather than the contaminated area.

The control of ventilation and air-conditioning equipment should be designed to meet the requirements of start-stop sequence and linkage control of various pieces of equipment.

1.5.3 Lighting System

Clean dust-proof lighting fixtures with enclosed covers that are easy to clean, rather than grille lights, should be provided at the medical facilities. The lights should be mounted to the ceiling, with their installation gaps sealed tightly.

Night lighting should be provided in the wards and through corridors, and the unified control of those lights from the nurse station is advisable. The light switches in the wards should be of wide-panel push-button switch, preferably mounted 1.2 m above the ground, to provide convenience for the elderly people.

Sockets for UV disinfection lamps or air sterilizers should be installed in medical facilities and other places requiring sterilization. Special switches with special marks and signal indicators, preferably mounted 1.8 m above the ground, should be provided for UV disinfection lamps, and shall not be in parallel with switches for ordinary lights.

The emergency evacuation lighting system should comply with the requirements of applicable fire protection codes.

Safe lighting, with an illumination 100% of that of ordinary lighting, should be provided in the operating rooms, emergency rooms and ICUs. Emergency lights should be installed in key medical equipment rooms and medical facilities.

1.5.4 Selection and Laying of Lines

When laid indoors, the wires and cables should be of low-smoke halogen-free flame-retardant type for ordinary load, while fire-resistant type or low-smoke halogen-free flame-retardant type for fire load. Armored cables should be used if laid outdoors.

Trunking and conduits may be laid in an exposed way, and should be non-combustible. The openings and gaps on the wall through which they run should be tightly sealed. When they cross the interfaces between the contaminated area, the semi-contaminated area and the hygienic area, the gaps and orifices of the trunking and conduits on the partition wall should be sealed tightly with non-combustible materials to prevent cross-infection.

1.5.5 Lightning Protection and Grounding System

Lightning protection and grounding of the makeshift hospital for infectious diseases should be designed according to the currently applicable national codes.

The low-voltage incoming power source should be grounded iteratively when entering the house, and the TN-S system should be adopted in the building.

A grounding system is shared for lightning grounding, protective grounding, functional grounding, shield grounding, anti-static grounding and so on.

The building is provided with main equipotential bonding. Auxiliary equipotential bonding should be adopted locally in the ICUs, operating rooms, emergency rooms, treatment rooms, bathrooms or restrooms with showers.

1.5.6 Communication, Network and Generic Cabling Systems

The computer system consists of internal network, external network and special equipment network, with redundancy in network architecture via physical isolation. The core switches of internal and external networks should be connected via jumpers to realize information communication.

Internal and external networks, points of information for voice and IPTV should be provided in wards, nurse stations, offices, conference rooms, medical facilities and dormitories. Wi-Fi should be accessible to the whole hospital area.

The network access points are advisably reserved for all the laboratory equipment rooms, testing rooms and reporting rooms at the medical test and imaging department. The network and telephone access points should be reserved at the sewage treatment station for online monitoring of water quality.

The network computer room should meet the requirements for installation of cabinets necessary for the information-based system, and be equipped with UPS, precision air conditioners, access control and environmental monitoring equipment.

1.5.7 Security and Protection System

Video surveillance systems should be provided at the entrances and exits of the hospital, external roads, and public areas of buildings, waiting rooms, nurse stations, and corridors.

Access control points should be provided at the entrances and exits of the ward area, the patient-doctor passages of the negative pressure ward, the transition area between the contaminated and hygienic areas, and the buffer rooms of the ICUs and the negative pressure laboratories, and meet the interlock control requirements of doors A and B.

The nurse stations and doctors' offices should be equipped with the one-button alarm system. Audible and visible alarm signals should be transmitted to the monitoring room.

In the negative pressure isolation ward, monitoring devices should be installed at the doors of the ward and the buffer room at the corridor side to monitor the differential pressure between the ward and the buffer room, and between the buffer room and the corridor. These devices should initiate an audible and visible alarm in case of out-of-balance differential pressure.

The access barrier with an automatic license plate recognition system should be provided at the vehicle entrance of the isolation test and imaging area, and the turnstiles with identity recognition system should be provided at the pedestrian entrance.

1.5.8 Automatic Fire Alarm and Broadcasting System

The automatic fire alarm and fire linkage system should be designed to meet the requirements of currently applicable national codes.

The fire emergency broadcasting system and the public address system should be shared.

1.5.9 Call Signaling System

The isolation ward area should be equipped with two-way visual intercom systems between the patient entrance and the reception room, and two-way intercom call systems between the wards and the nurse station.

The operating area should be equipped with two-way intercom call systems between the nurse station and operating rooms. The care units, observation rooms and other medical facilities should be equipped with two-way intercom call systems between the nurse station and ward beds.

A one-way intercom system should be provided between the control room and the radiological apparatus room of the radiology department.

The ICUs should be equipped with mobile remote visitation systems, with their signals accessible via wireless access points.

The queue calling systems should be provided at the triage tables of the outpatient department and medical test and imaging department.

1.6 Air Conditioning and Ventilation Design to Control Pollution and Improve the Environment

1.6.1 Design Principles

Air conditioning (AC) systems were provided in all occupied rooms, and its outdoor parameters were given based on the winter meteorological conditions in Wuhan during the epidemic outbreak. Mechanical ventilation systems were used in ward areas, medical staff areas, and test and imaging areas. The mechanical AC systems in hygienic areas, semi-contaminated areas, and contaminated areas were operated independently. For each system, the supply fan and the exhaust fan were interlocked and placed in hygienic areas.

1.6.2 Air Conditioning System

The design of AC systems and its operation modes are mainly determined by the partitioning strategy and the specific function of each area. To prevent cross infection, AC systems were operated independently in contaminated areas, semi-contaminated areas, and hygienic areas. Direct expansion (DX) all-air purified air handling units (AHU) were used in area operating under negative pressure (NP) environment, including NP ICUs, NP test centers and NP operating rooms with AHUs being placed in specific equipment rooms. Split type heat pump air-conditioners were used in other areas. Due to a supply problem, electric heaters were adopted and served as the heat source for fresh air system of split type heat pump air-conditioners. Condensed water generated by AC systems was collected individually in each area and was then treated with sewage system and waste-water system.

1.6.3 Ventilation System

Air flows from the hygienic area to the semi-contaminated area then to the contaminated area, and the differential pressure between adjacent connected rooms with different pollution levels is not less than 5 Pa. The degree of negative pressure decreases gradually from the ward restroom, ward room, to the buffer room and the semi-contaminated corridor. The air pressure in the hygienic area shall be kept at a positive value against the air pressure outdoor. The air supply system is installed with three-stage (G2 low efficiency, F7 medium efficiency, and H13 high efficiency) filters, and the exhaust system is installed with H13 high efficiency filters (except for the medical staff area).

The ventilation frequency of a negative pressure isolation ward is not less than 12 times/h, and the exhaust volume is equal to the make-up air volume and infiltration air volume maintaining the room pressure. The ventilation frequency in the medical staff corridor and buffer test and imaging area is 6 times/h, and the exhaust volume is equal to the air volume necessary to maintain the pressure value. The ventilation frequency in the buffer room is 6 times/h, and the exhaust volume is determined according to the balance between the air supply volume in the buffer room and the volume of air infiltrating from the door gap.

The air supply and exhaust system for negative pressure isolation wards should be set up in a collective manner, and every 6 rooms and their restrooms share one set of air supply and exhaust system. The ward has an upper air outlet, and a lower air outlet near the bed head to ensure unidirectional air flow and protect the safety and health of medical staff (Figure 1.34).

1.6.4　Pipeline and Equipment Layout

The air supply and exhaust pipes of the negative pressure isolation ward and restroom directly enter the room from the side wall. The wards looks concise since no transverse air duct is installed. The fan inlet is installed with an electric closed air valve linked with the fan. All air supply and exhaust branch pipes are installed with constant-air-volume air valves, and the air supply and exhaust branch pipes in each ward are installed with electric closed valves that can be turned on/off independently. A differential pressure gauge is mounted on the wall along the ward and medical staff corridor, to show the differential pressure in different areas for medical staff and maintainers to observe the pressure gradient of a room in real time and judge

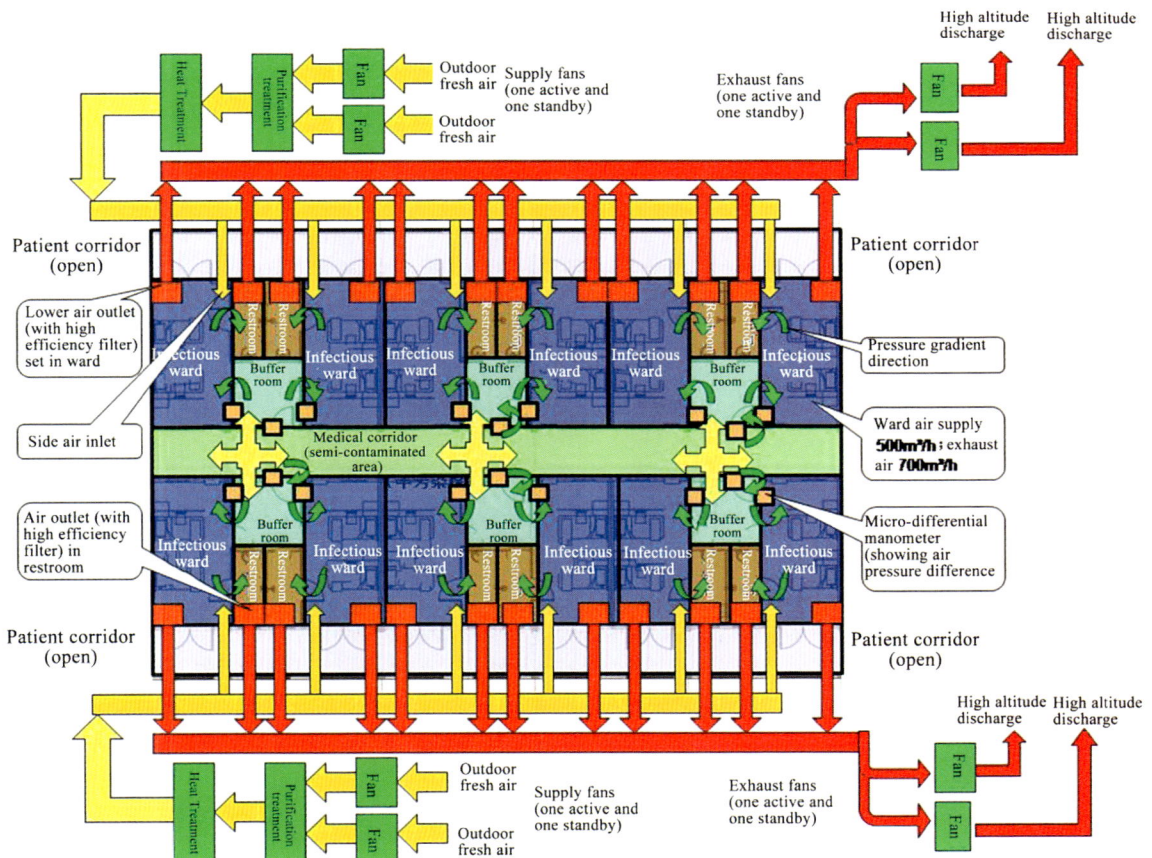

Figure 1.34　Diagram of Ventilation System in Ward Area

whether the air supply and exhaust system is operating normally.

1.6.5 Air Distribution Simulation in Ward

During the design process, Central-South Architectural Design Institute Co., Ltd., in cooperation with Dassault Systems (Shanghai) Information Technology Co., Ltd., carried out airflow distribution and contaminant concentration simulation, and compared and analyzed the effects of different air supply and exhaust schemes on indoor airflow and contaminants with the transient XFlow software. The simulation result indicated that, the airflow distribution of upper air supply and lower air exhaust can form a reflux test and imaging area at the sickbed, thus effectively removing polluted air from the wards in time.

1.6.6 Indoor Temperature Forecast for Wards

While considering the winter heating plan, the design team made a forecasting analysis on the indoor temperature of wards in the upcoming cooling season after the heating season ends. To meet the design temperature requirement of 26°C in the cooling season, the air supply system must be renovated accordingly in June, and the measures are as follows： ① maintaining the original air supply volume of the room and adding refrigeration equipment to the air supply system； ② adding no refrigeration equipment to the air supply system, and reducing the air supply volume to 550 m^3/h. After June, refrigeration equipment must be added to the air supply system of the wards.

1.6.7 Numerical Simulation of Exhaust Emission

Professor Lu Xinzheng's team from Tsinghua University, based on FDS (an open source fluid mechanics computing program), proposed a method for rapid simulation of the environmental impacts caused by exhaust from temporary hospitals. This method, based on the distributed computing of cloud computing platform, monitoring and visualization of polluted air flow, provides a special tool for rapid analysis of temporary hospitals in the design. In the design, exhaust air from ward areas is discharged from the roof at a height of 6.0 m. The fresh air inlet and outlet on roof are spaced at a horizontal distance of 20 m and a vertical distance of 3.0 m. Through simulation, it is believed that this solution can effectively avoid mixing of inlet and outlet air, and the pollution of exhaust to outdoor environment.

1.6.8 Indoor Environment Monitoring

Professor Lin Borong's team from Tsinghua University has carried out real-time, high-density, and high-accuracy monitoring over the indoor CO_2 concentration, $PM_{2.5}$ concentration, temperature, relative humidity and other parameters for areas where patients and medical staff stay. Over 24-hour monitoring results of environmental parameters for typical wards shows that, the CO_2 concentration fluctuates in the range of 520-670 ppm and peaks at 14：30 and 21：00, due to the increase of personnel density as medical staff enter for ward rounds. The concentration of $PM_{2.5}$ peaks at 11：00 and then fall back at 14：00. This indicates that, the contaminant concentration is high during this period, and medical staff faces a greater protection pressure. For the rest of the time, the concentration is relatively stable, and the concentration line tends to be straight. The room temperature fluctuates at 20-22 ℃, and the relative humidity fluctuates between 40%-55%. Monitoring data in major time periods meet the design requirements.

1.6.9　Selection of Equipment and Materials

If possible, the design scheme shall prioritize mature and reliable equipment donated from manufacturers with large inventory and fast transportation, so as to shorten the time of product procurement and allocation, and facilitate quick installation and simple commissioning. The selection of equipment and materials shall also meet the national codes.

1.6.10　Installation and Commissioning

It is important to ensure the air tightness of building envelope, and pay special attention to the waterproof and sealing treatment of such envelope where pipelines (air ducts, refrigerant pipes, etc.) penetrate the roof and side walls. Considering the requirements of construction schedule, the engineers use PE pipe featuring fast construction and satisfactory air tightness to replace conventional air duct for outdoor installation. Special attention should be paid to the pressure gradient in negative pressure rooms (quarantine wards, ICU, laboratories, operating rooms, etc.), and ensure that the indoor pressure gradient of each area meets the design and functional requirements.

1.6.11　Summary and Reflection

"Three areas and dual passages" is an important design concept for emergency hospitals, which helps control a reasonable pressure gradient and ensure directed airflow. It is not recommended to design a complicated automatic control system, but it is advisable to use easy and practical means to meet relevant functional requirements. It is suggested that backup measures should be taken for ventilation equipment in negative pressure isolation wards, and other life support equipment should feature redundant design. An important link for such engineering construction is to find and solve problems in time during follow-up field services and drawing review, and ensure onsite guidance and coordination by designers.

1.7　Project Cost Analysis

1.7.1　Analysis of Cost Differences between Project Design and Conventional Hospitals

As the hospital is a temporary emergency project, and its function demands clearly focus on diagnosis and treatment of severe patients, the outpatient area is canceled in the design.Instead, 59 ICU beds, 400 m^2 negative pressure laboratory, and 3 radiation-proof CT rooms are provided. The disinfecting area occupies a relatively large proportion in the plane layout. All wards are equipped with UV disinfection delivery windows to reduce contact. A buffer room is set at the entrance of a ward, with full suspended ceilings to improve the negative pressure environment. In order to meet the urgent demand of temporary emergency project, we choose the light steel modular buildings as the main body of the Project, and select the light steel sandwich panels as the main body of the test and imaging area with high floor height requirements and high bearing capacity.

The power supply load class of the project is determined as class 1. Four 10 kV HV power supplies are introduced from different regional substations, making the overall power supply & distribution capacity of the hospital more than twice that of ordinary hospitals. UV disinfection lamps are provided in wards, restrooms, corridors, consulting rooms, operating rooms, and other venues requiring disinfection. Considering the provisionality of the project, the Tel+IDF intelligence only involves the basic automatic fire alarm and linkage control, integrated wiring, security, and ward call.

To prevent the water backflow from polluting the municipal water supply network, the project incorporates the break tank with pressurized feed pump station for water supply. The outlet pipe of the pump room is provided with a UV disinfector, and emergency chlorination measures are taken for domestic water supply to ensure the safety of water supply. For prevention of cross infection in the isolation ward and area treatment, three independent outdoor drainage networks are provided to treat the sewage from the ward and treatment area, and other areas, and outdoor rainwater

respectively. Sewage from the isolation ward and treatment area, after being collected by the independent pipe network, will enter the sewage treatment plant for unified treatment and up-to-standard discharge. HDPE impermeable films are paved on the floor outside the isolation ward and treatment area, to prevent contaminated rainwater from seeping into the ground, resulting in groundwater pollution. Due to the tight schedule of the project, fire hydrants and automatic sprinkler system are not provided. Instead, portable fire extinguishers are provided as per the hazard level to ensure that portable fire extinguishers cover the 15 m protection range. For prevention of external infection due to contaminated water, sewage from the ward and treatment area will go through the process of "pre-disinfection + septic tank + secondary treatment + disinfection" with a longer water retention time. The sewage treatment plant has a total capacity of 80 m^3/h, several times over the capacity of regular hospitals.

The project requires forming a gradient differential pressure from the hygienic area, semi-contaminated area, to contaminated area through air supply and exhaust distribution, to control the overall direction of air flow in the infectious disease ward and treatment area. A mechanical ventilation system shall be set up to cover the entire ward and treatment area. Constant-air-volume valves are installed on the air supply & exhaust branch pipes of the ventilation system in all areas, and electric closed valves are installed on both the air supply and exhaust branch pipes in every ward. For the reduction of bacteria and virus concentration in wards, the ventilation rate in wards for respiratory infectious diseases must be twice (or more) as much as that of the wards for non-respiratory infectious diseases, and the capacity of the air supply/exhaust fans must increase accordingly. All fans work in one-duty and one-standby mode, and each air supply system is equipped with three-stage filter (low, medium, and high-efficiency filtration). Exhaust air system is provided with high-efficiency filtration. According to function demand, a medical gas and pipe network system is set up for the project to provide oxygen, negative pressure suction, and compressed air.

1.7.2　Analysis on Cost Differences in Project Construction Organization

I. Organization and Management of Workers

Due to the project's tight schedule and large scale (about 80 000 m^2), the construction of Leishenshan Hospital requires a large number of workers. However, the peak period of the Project coincides with the epidemic control period and Spring Festival, so staffing organization is a big challenge. Since the extremely short construction duration makes it impossible to use the conventional flow construction with steps, lots of cross operations lead to large reduction in work efficiency and great increase in labor costs.

II. Supply and Organization of Equipment & Materials

The project proceeds during the national epidemic control period, when most factories are shutdown, and market supply relies on existing stocks. Moreover, owing to the tight schedule and goods transportation difficulties (since the procurement scope of materials and equipment covers the whole country), the price of project equipment and materials rise sharply. Some out-of-stock non-standard equipment needs to be customized, and non-full-load operation of the factory leads to sharp increase in equipment cost.

III. Supply and Organization of Construction Machinery

As affected by the Spring Festival, epidemic prevention, and traffic control, the machinery and equipment are substantially mobilized into the site at once and demobilized one after another as construction needs, which makes the mechanical efficiency drop obviously. For example, a large number of various cranes are mobilized at the early stage, to cope with the centralized hoisting of numerous containers and roof ventilators. However, the work intensity of the early stage is relatively low. After the container houses are substantially transformed at the site, cranes are used for centralized hoisting in large batches, with extremely strong work intensity. So the efficiency of construction machinery drops dramatically.

IV. Costs of Management and Measures

Due to the extremely tight schedule of the Project and a large number of cross operations for field construction, the number of staff engaged in material allocation, business finance, construction management, and quality management has increased by dozens of times over that of conventional projects, to ensure the progress of the project. The project proceeds during the peak of the epidemic and the site is densely populated, which increases the protection expenses for epidemic control, as well as the related expenses for quarantine of workers and management staff after construction.

1.7.3　Discussion on Project Cost

I. Analysis of Project Cost Characteristics

It can be seen from the aforementioned design and construction organization characteristics that, the incremental cost of respiratory emergency hospitals is much higher than that of conventional hospitals, and the incremental cost of emergency items is all-round compared with that of the conventional items. Emergency project follows a construction mode different from conventional projects. The primary objective of an emergency project is the progress, which the design and construction organizations also focus on. The design work often needs to be adjusted according to the existing resources, so economical efficiency is not a controlling factor for project implementation. Moreover, due to the tight schedule of the project, lack of pre-planning time, and insufficient measures for cost control, the execution efficiency of the project drops obviously but the cost increases.

II. Relevant Suggestions

Based on the analysis of the above cost characteristics, it is suggested to strengthen the emergency response and control capacities from several aspects, so as to reduce the incremental cost due to efficiency reduction.

（1）To strengthen the development and application of information technologies. Information technologies can effectively improve the supply and management capabilities of staff, materials, and machinery, optimize enterprise resources, and strengthen emergency

dispatching capabilities, so as to complete a project with more economical, reasonable, and effective allocation of staff, materials and machinery. Furthermore, the information technology application can effectively save the data generated during the project implementation and provide a strong basis for project settlement and auditing in the future.

（2）To enhance the BIM application capability. BIM application can effectively simulate the field construction organization, and provide solid references for actual construction organization. Enhancing this application capability can serve the project more timely and effectively, especially for the execution simulation of emergency projects. This can optimize the manpower allocation, construction flow organization, and improve work efficiency based on the simulation results.

1.7.4　Summary

To sum up, the emergency hospital project, especially the respiratory emergency hospital, shows prominent characteristics and many cost differences compared with the conventional hospital project, and is totally different in temporary emergency items. Therefore, during the early decision-making stage of the project, it is suggested to adopt the mode of "conventional cost index + incremental cost", which means the incremental cost is determined based on comprehensive consideration of the current labor, materials and market supply. During the implement of the project, it is suggested to settle accounts by means of cost plus remuneration, agree on the confirmation method for costs in advance in the contract, and strengthen the tracking and auditing supervision, thus laying a good foundation for the smooth progress and later settlement of the project.

2

Makeshift Hospital

2.1 Design Strategies for Construction of Wuchang Makeshift Hospital (Gymnasium)

In 2020, when people across the country are preparing for the Spring Festival, a sudden COVID-19 epidemic swept across the country, wreaking havoc in Hubei Province. The Wuchang Makeshift Hospital was rapidly built to prevent the spreading of the virus. It was set up to tackle the disease, by means of alleviating the shortage of beds and medical resources in short time. A three-dimensional prevention and control system was formed with Leishenshan Hospital and Huoshenshan Hospital, designated hospitals, and other medical institutions, to quickly get the epidemic under control. Wuchang Makeshift Hospital was one of the first of the three makeshift hospitals established in Wuhan, and was put into use on February 5. The development of Wuchang Makeshift Hospital only took 48h from decision-making to admitting the first batch of patients.

2.1.1 Concept Analysis of Makeshift Hospital

Since the release of announcement that Wuhan would be under lockdown, the number of confirmed patients in Wuhan had been soaring. In order to cope with the epidemic, the government draw up the coping strategy of "leaving no one unattended". Academician Wang Chen proposed the concept of makeshift hospital : setting up beds and ward and treatment areas in existing public facilities ; transforming large scale public buildings in the city ; and rapidly building medical shelters with sheltered spaces and "three areas and dual passages".

2.1.2 Construction Principles

I. Rapid Construction

The construction of a large emergency hospital normally takes at least two years, while the Wuchang Makeshift Hospital was completed in two days.

II. Massive Scale

Hongshan Gymnasium can accommodate up to 800 beds, just to meet the demand of massive patients with mild symptoms.

III. Low Cost

The renovation of Wuchang Makeshift Hospital mainly involves the division of "three areas and dual passages", and no large amount of civil works is required. Compared with the construction cost of the emergency hospitals of the same scale, the conversion cost of Wuchang Makeshift Hospital is a drop in the bucket.

IV. Essential Functions

Essential Functions include isolation, triage, medical treatment, referral, essential living and social engagement.

2.1.3 Design Strategies

I. Site Selection

Hongshan Gymnasium is located in the center of Wuchang, with accessible transportation, open view, high altitude, flat ground and good land use conditions.

II. Rapid Construction

1. Selection of Functions

The conversion of Hongshan Gymnasium focuses on the principle of "three areas and dual passages". The original building was converted into the isolation ward area and logistics area. The functional units to be added include the safe passage unit, medical treatment unit, sewage treatment unit, unit of domestic water for patients, and triage unit.

2. Selection of Construction Materials

The partition walls are made of color steel laminboards, which have good performance and features that are convenient for rapid construction. The integrated restrooms, septic tanks and finished wash basins are used for field assembly and placement. Tents are used to set up outdoor triage rooms.

III. Meeting Functional Requirements

1. General Layout

According to functional division, the hospital can be divided into isolation area, living area for medical staff, triage and treatment area and sewage treatment area.

Entrances and exits are properly set according to functional requirements. Flows for hygienic and contaminated materials shall be separated.

2. Setting of Building Functions

The "three areas and dual passages" principle is the core for the renovation of the makeshift hospitals. Hongshan Gymnasium was used as a sports building before. Though detailed field investigation, it basically meets the conditions for the renovation of the makeshift hospital. The building is strictly divided into the hygienic area, contaminated area, semi-contaminated area, passageway of medical staff and passageway of patients.

3. Equipment Guarantee Proposal

（1）Water Supply & Drainage Design.

According to the functional division of the building, venues for medical staff is provided with fixed toilets and shower stalls, and wash basins are provided in the nurse station and changing room for medical staff. Patients use temporary mobile toilets, wash basins and temporary showers outside the venue. Sewage produced by medical staff is collected by the existing sewage pipeline system in the venue, and sewage produced by patients is collected by the newly-built outdoor overhead pipeline system. The two sewage collection systems are completely independent from each other.

（2）Ventilation Design.

For renovation of the air conditioning and ventilation system, operators make full use of the original system, facilities, and equipment as far as possible in a fast and easy way. The three ward and treatment unit systems are renovated to maintain negative pressure for the ward and treatment units. Air conditioning systems in the hygienic area and contaminated area operate independently. Exhaust air from the contaminated area and semi-contaminated area must be treated before discharge. In the hygienic area and ward area, several sets of air disinfecting machines with the sterilization function are set up to purify the indoor air.

（3）Electrical Design.

For electrical transformation of Hongshan Gymnasium, the team focuses on distribution safety and considers the rationality and ease of transformation. While meeting the electricity demand of the hospital, the design also con-siders the convenience of equipment operation and maintenance.

2.1.4　Precautions and Promotion Strategies for Gymnasium Conversion

It needs to meet three prerequisites for building conversion : gross volume, large space, and few floors.

What should be carefully dealt with are the fire compartment and partition settings of the original building and absolute separation of hygienic and contaminated materials for safety.

The number, location and area of the safe passways should be set reasonably to avoid the problems, such as too far away and too long waiting time. Take care of patients' mental health, reserve enough activity areas, and organize social activities.

2.1.5　Summary

On March 10, 2020, 5 : 00 p.m., Wuchang Makeshift Hospital in Wuhan, Hubei Province was officially closed down. Makeshift hospital, a special hospital that played a key role in the prevention and control of the COVID-19 epidemic in Wuhan, has successfully completed its historical mission.

2.2　Design of Water Supply, Drainage and Fire System for Makeshift Hospital— A Case Study on the Sports Center's Ping-Pong & Badminton Gym Conversion Project

2.2.1　Project Overview

The Ping-Pong & Badminton Gym of Jiangxia Dahuashan Outdoor Sports Center was completed and put into use in August 2019. It has completed indoor and outdoor water supply, drainage and fire systems. To serve COVID-19 patients with mild symptoms, the Ping-pong Hall on the first floor and Badminton Hall on the second floor (west area) of the Ping-Pong & Badminton Gym were converted into a makeshift hospital house with a total of 628 beds.

2.2.2　Domestic Water Supply System

The maximum domestic water consumption of the makeshift hospital is 198.4 m^3/d. The water supply pipeline comes from the municipal water supply network, and a reduced-pressure type backflow preventer is set to prevent backflow contamination.

All wash basins are equipped with induction faucets.

2.2.3　Domestic Hot Water and Drinking Water System

I. Hot Water System

Domestic hot water is available in showers in the living area for medical staff, of which all are equipped with an individual electric water heater with a heat storage capacity of 80 L.

II. Drinking Water System

Each care unit is provided with a separate drinking water supply point and a 12 kW electric water boiler with built-in filtering device. The drinking room should be located close to the restroom, so that water can be discharged into the floor drain nearby.

2.2.4　Sewerage System

I. Sewage Collection and Disinfection

The maximum water discharge of the project is 198.4 m^3/d. The domestic sewage produced by medical staff and that from the isolation area are discharged separately.

Medical staff's domestic sewage is discharged to the existing outdoor sewage inspection well, and disinfectants are added to the septic tank for primary disinfection. The domestic sewage from the isolation area flows through the gravity pipe to the sewage tank, where primary disinfection is applied to the water inlet. Sewage, after staying in the sewage tank for at least 1.5 h, is discharged to the septic tank, where it combines with the domestic sewage from medical staff for secondary disinfection. Finished septic tank is adopted for the sewage tank. The sewage tank and septic tank are added with 10% chloros solution by automatic dosing, and the chloros solution is prepared on site with finished chloros (Figure 2.1, Figure 2.2). Sewage should stay in the disinfecting tank for no less than 1.5 h, in which the residual chlorine should be over 6.5 mg/L (by free chlorine) and fecal coliform less than 100 mg/L. The reference effective chlorine dosage is 50 mg/L.

II. Setting and Disinfection of Vent Pipes

Restrooms uses the existing vent pipes, which have been fixed to ensure the fresh air inlets are far away from the vent pipe outlets. Facilities (mobile toilets, sewage tanks, and septic tanks) have built-in vent pipes, which shall be set in a well-ventilated place without affecting the passage. Vent pipes of such facilities have their pipe openings disinfected by

Figure 2.1　Sewage Tanks in Makeshift Hospital

Figure 2.2　Sewage Disinfection and Dosing Equipment in Makeshift Hospital

Figure 2.3　Mobile Toilets in Makeshift Hospital

ultraviolet disinfection equipment (Figure 2.3) .

2.2.5　Fire System

This project uses the original firefighting facilities and takes measures to ensure they will not be shielded. For fire safety of the project, fire extinguishers are arranged at closer interval on the basis of danger level. Wheeled fire extinguishers MFT/ABC (20kg, 6A) are provided for large spaces (such as the gym) at a protection distance of 20 m, and portable fire extinguishers MF/ABC (5kg, 3A) are provided for other medical office areas at a protection distance of 10 m.

2.2.6 Pipe Selection and Laying

Propylene random copolymer (PPR) pipes with hot melt connection are used indoors and outdoors in domestic water supply. Outdoor water supply pipes are exposed in the green belt. Indoor water supply pipes are exposed on the wall without diameter reduction and pipe specification is lowered to facilitate procurement and construction.

Indoor and outdoor mobile toilets are all built on stilts and pipes are exposed on the ground to minimize pipeline excavation. Polyethylene (PE) pipes with hot melt connection are selected for discharge.

2.2.7 Summary

I. Domestic Water Supply System

For the selection of water supply system, it is suggested that the break tank should be used when conditions permit and the outdoor integrated water pumping station is preferred for easy construction and installation. Moreover, an emergency water replenishing port and emergency chlorination port are reserved on the water tank, to ensure the water quality and quantity.

II. Domestic Hot Water System

For domestic hot water system, it is suggested that an air-source heat pump centralized hot water system should be set up if conditions permit; or otherwise, an electric water heater centralized hot water system should be established.

III. Domestic Sewerage System

Sewage from the medical staff area and isolation area must be separated and subject to secondary and primary disinfection respectively. The existing septic tank will be utilized as the disinfecting tank.

IV. Disinfection

For disinfection of vent pipes for pipelines and facilities, it is recommended to use ultraviolet disinfection equipment and high-efficiency filters are not recommended.

For sewage disinfection, it is recommended to use finished chlorine dioxide or finished chloros as the disinfectant, instead of liquid chlorine, chlorine dioxide and chloros requiring field preparation.

V. Fire System

For venues with large space, it is recommended to provide wheeled fire extinguishers.

VI. Pipe Selection and Laying

It is recommended to use plastic pipes for drainage, which should be of hot melting or electric melting connection, instead of socket-and-spigot connection.

Exposed laying with proper protective measures is recommended for water supply and drainage pipes.

2.3 Key Points of Electrical Design

This section outlines the electrical design of the "China Optical Valley" Rihai Makeshift Hospital converted from factory. During the conversion, it is advisable to conduct comprehensive site survey, bearing the function and easy construction in mind to make reasonable innovation and improvement without violating the provisions of the codes.

The "China Optical Valley" Rihai Makeshift Hospital was composed of 4 mobile cabins and boasted the largest number of beds among the makeshift hospitals in Wuhan. The mobile cabin B (former metal plate workshop) has a conversion area of about 9 984 m^2 and holds 610 beds. The mobile cabin C (former machine room division) has a conversion area of about 16 156 m^2 and holds 1 445 beds. The mobile cabin D (former injection molding workshop) has a conversion area of about 8 319 m^2 and holds 654 beds. The mobile cabin E (former tower manufacturing department) has a conversion area of about 12 081 m^2 and holds 980 beds. They all are single-storey buildings.

2.3.1　Power Supply and Distribution Design

Rihai Makeshift Hospital is a temporary medical building and an isolation ward and treatment area for infectious diseases. Therefore, it should comply with the relevant codes of healthcare facilities and emergency hospitals. However, it needs to be completed and put into use within a very short period. To comply with related codes while satisfying requirements on construction period and equipment purchase, the electrical system should be simple and reliable and the equipment should be easy to purchase and install.

According to the *Code for Electrical Design of Healthcare Facilities* and the *Code for Design of Infectious Diseases Hospital*, in combination with the use requirements, the loads of the power supply equipment involved in the conversion are classified as follows (Table 2.1) .

Load Class and Description of Electric Equipment　Table 2.1

Load Class	Description of Electric Equipment
Especially Important Load in Class I Load	Emergency lighting and evacuation indication system; Electricity consumption in surveillance center and intelligent system room; Positive pressure blower in medical staff area
Class I	Medical equipment and lighting at the treatment room, nurse station, etc.
Class II	Electric diagnostic equipment （CT, etc.）
Class III	Ordinary air conditioner and some equipment under other loads excluding Classes I and II loads

Notes: Loads in the above table are classified according to the engineering construction standards in China.

During the conversion, an 800 kW outdoor diesel generator is provided to supply power under the Class I load and the especially important load in Class I load. The diesel generator has its own fuel tank and oil supply port. In order to reduce the adjustment and modification of the original power supply system and shorten the construction period, the diesel generator is integrated into the mains bus for power supply. The outlet switch for such a diesel generator is interlocked with the outlet and interconnection switches of the transformer. Only when the outlet switch for transformer is off can the outlet and inlet switches for diesel generator be used.

For the power supply under different load classes, it is advisable to consider the electricity demand and the actual site conditions, take the existing power distribution cabinets on the site as the local master box to reduce the dismantling and modification of the original system and the additional provision of field equipment. The especially important load in Class I load is switched and powered by the mains supply and diesel generator in the power transformation and distribution room, and the Tel+IDF equipment is provided with UPS additionally. Class

I load is switched and powered by the mains supply and diesel generator in the distribution room. Classes II and III loads are powered by mains supply. Due to the connection lines provided in the power transformation and substation room, the Classes I and II loads can also achieve the function of dual mains supply.

2.3.2 Lighting System Design

The illuminance refers to the high standard value stipulated in the *Standard for Lighting Design of Buildings* and meets the provisions on the target value of load density. The decontamination area for medical staff is illuminated by ceiling-mounted clean luminaires with closed covers that are difficult to accumulate dust and easy to wipe. UV disinfection lamps are provided in changing rooms, shower stalls, and other places with independent switches. The ceiling-mounted fluorescent lights with transparent covers and the night lighting should be provided in the isolation ward area. Considering that the power supply radius should not be too long for convenient centralized management, lighting distribution boxes should be provided in the medical staff area and ward area respectively.

Because of difficulties in the source of goods, the non-centralized control evacuation lighting system, rather than the centralized one, is adopted. Luminaires with built-in storage batteries, of which the lighting can last no less than 60 minutes, are adopted. In case of fire, the main power output of the distribution box for emergency lighting can be cut off manually, and all the non-continuous emer-

gency lamps can be lit at the same time. The light source of the continuous lamps changes from the power-saving mode to the emergency lighting mode.

2.3.3 Information Access and Network System Design

I. Information Access

In order to save the on-site construction time, the equipment has been utilized for information access in the original plant area. Four optical cables are routed by the telecom service provider into the mobile cabins B—E respectively, and connected to the local computer network at the Tel+IDF room. Because the hospital requires internal communication, the mobile cabins are interconnected by 12-core optic cables. The bandwidth of each cabin exit is 500M, which can meet the demands of patients and medical staff.

II. Telephone System

(1)Setting of telephone information socket: One voice point is reserved every 5–10 m^2 at the nurse station.

(2) Network access method: The telephone is accessed via the virtual network without a program-controlled switch. The telephone access device is arranged in the surveillance room of makeshift hospitals and accessed by the operator via optical cables.

III. Computer Network

This project is designed to have two physically-isolated networks, namely the computer information network and computer security network.

The information network realizes the logical isolation of internal and external networks by dividing the VLAN network segment. The internal network is for the communication of medical staff and equipment. The external network carries Internet information for use by medical staff, administrative staff and in-patients. The security protection network is used to carry the Tel+IDF communication signals of the security monitoring system.

IV. Wi-Fi Network

High-density wireless APs are installed in treatment areas, duty rooms, offices, nurse stations and corridors to achieve a fully-covered wireless network. POE is used for wireless AP. AP is set with two SSIDs. The SSID for patients only provides the access service for external network while the one for medical staff provides the access service for both internal and external networks.

The network speed in the treatment area is limited to 400 M by management software while that in the working area is limited to 100 M. Priorities are given to the medical staff in the working area.

2.3.4　Security Protection System

Every unit in the treatment area has one monitoring point to monitor the conditions in the cabin and panoramic cameras are installed high in main corridors. According to actual site conditions and requirements of the public security department, the installation height should be above 6.0 m to have a comprehensive view of the cabin.

After heavy work, the exhausted medical staff often make careless mistakes when tak-ing off the protection suit. Therefore, a dome camera is directly installed above the mirror in the room where the medical staff take off the isolation suit and protection suit, allowing the protection supervision team to check whether the standard requirements and the correct sequence are followed. In case of any wrong action or omission, the team will give timely reminders and take remedial measures by broadcasting to the room, so as to ensure the safety of the medical staff.

2.4　Discussion on the Renovation of Ventilation and Air Conditioning of Makeshift Hospitals from the Large Space for COVID-19

2.4.1　Introduction

For prevention of virus spreading, the existing large space (gymnasium, exhibition hall, factory, etc.) are converted into makeshift hospitals to receive and cure confirmed patients with mild symptoms of coronavirus. This section discusses the method of reasonably converting the contaminated area and semi-contaminated area in a short period of time, which can not only protect the health of medical staff and patients, but also prevent virus from spreading into the surrounding environment.

2.4.2　Characteristics of Buildings to be Converted and Requirement for Conversion

I. Characteristics of Buildings to be Converted

The existing large spaced buildings show the following common characteristics : ① high and large space, numerous doors and windows, and poor air tightness; ② with central air conditioning system.

II. Ventilation and Air Conditioning Facilities

Generally, the terminal mode of air conditioning incorporates a primary return air all-air AC system with a mechanical exhaust system, whose exhaust air volume is less than the air supply volume of AC unit.

III. Renovation Requirements and Characteristics of Ventilation and Air Conditioning

The conversion requires a very short time limit and ready-to-use equipment and materials based on the situation. When they are closed, the makeshift hospitals should be able to resume their original functions very quickly.

2.4.3 Analysis on Ventilation and Air Conditioning Renovation of Makeshift Hospitals

I. Ward and Treatment Area

1. Sickbed Layout

As the outdoor temperature is relatively low in winter and spring, the area near the external window will form a wall jet, which destroys the heat plume on the surface of human body, making it easy to cause cross infection. Therefore, sickbeds shall be kept away from this area, which can also protect patients from the cold radiation.

2. Impact of Supplied and Exhausted Air Volume on Indoor Negative Pressure in Large Space

The hygienic area, semi-contaminated area, and contaminated area need to maintain an orderly pressure gradient. The ward and

treatment area has the lowest pressure, but large space has poor air tightness. For prevention of virus overflow, it is preferred to choose the venue with suitable height and fewer exterior windows to convert into a makeshift hospital. If a building shows poor air tightness, the air supply volume shall be reduced to maintain the indoor negative pressure. It is recommended to take 60%–90% of the exhaust volume as the air supply volume. Take the lower value for poor air tightness. Otherwise, take the higher value.

3. Conversion Method of Ventilation and Air Conditioning

Converting the terminal AC system to an all-fresh air operation mode involves the following steps: ① to close the return air valve and turn on the fresh air valve to VWO state; ② to add an exhaust filter group; ③ to adjust the fresh air volume to keep the indoor negative pressure. These steps are simple, reliable, less expensive and of less chance of infection in O&M.

4. Indoor Temperature in All-Fresh Air Operation Mode

Indoor temperature has a great impact on patients. The appropriate indoor temperature is 16–28℃. When the indoor temperature doesn't reach the optimal value, other temporary measures can be adopted. In case of winter, these temporary measures may include: ① increasing the flow of hot water; ② raising the temperature of supplied water.

5. Exhaust Air Volume per Capita

The exhaust air volume needs to comply with following conditions: ① meeting the patients' demand on fresh air volume; ② establishing an orderly pressuring gradient; ③ keeping appropriate indoor temperature; ④ facilitating the rapid conversion. It is suggested that the exhaust air volume be not less than 90 m³/h per person. For areas with suitable local temperature during the epidemic period, the exhaust air volume can be appropriately increased.

II. The Entrance and Exit for Medical Staff

The entrance and exit for medical staff are the key areas that need to be highlighted for prevention of cross infection. The diagrams of a typical entrance and exit for medical staff and ventilation layout are shown in Figure 2.4 and Figure 2.5.

1. The Entrance for Medical Staff

It is reasonable to maintain positive pressure for changing room 1 with air supply for the purpose of preventing virus from spreading

to the hygienic area. *DN*300 short ventilation pipes are used for adjacent compartments and proper supply fans are selected based on the air supply volume to changing room 1 (30 times/h or 1 293 m³/h, whichever bigger). The supply fan shall be a constant frequency blower with a gentle performance curve. A manual closed air valve is set on the short pipe between changing room 2 and the buffer room. The buffer room and the ward and treatment area make use of short pipes with electric air valves, which are interlocked with the supply fan.

2. The Exit for Medical Staff

The exit for medical staff should have the following characteristics: ① it takes extremely short time for medical staff to pass through the buffer room; ② medical staff stay longer in the isolation suit take-off room where the bacteria concentration may reach the summit, and the isolation suit and mask may be stained with

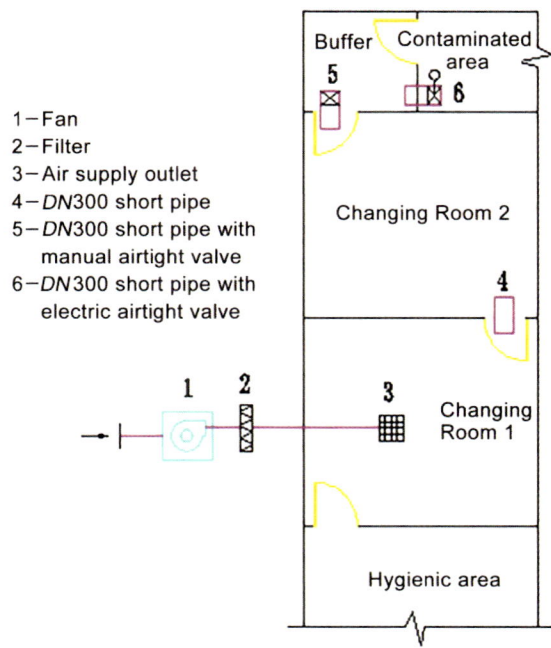

Figure 2.4   The Entrance for Medical Staff

1—Fan
2—Filter
3—Air supply outlet
4—DN300 short pipe
5—DN300 short pipe with manual airtight valve
6—DN300 short pipe with electric airtight valve

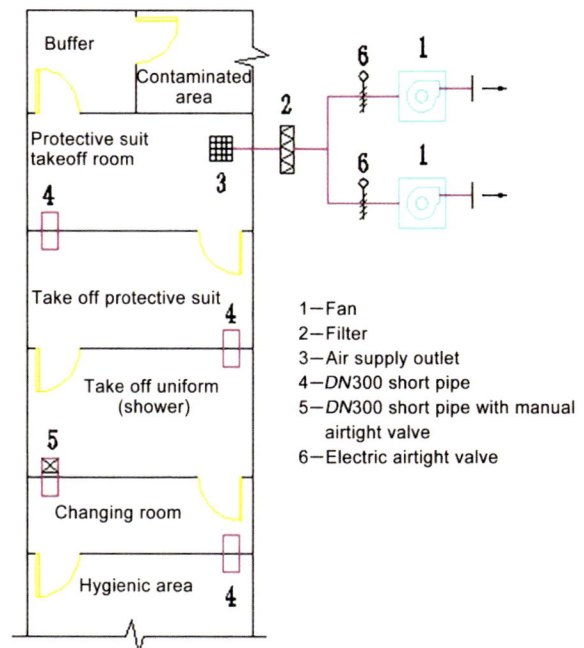

Figure 2.5   The Exit for Medical Staff

1—Fan
2—Filter
3—Air supply outlet
4—DN300 short pipe
5—DN300 short pipe with manual airtight valve
6—Electric airtight valve

virus again. A reasonable pressure gradient can be established according to the air exhaust solution shown in Figure 2.5. The air volume of the exhaust fan in the isolation suit take-off room takes the larger value with references of 30 times/h and 976 m^3/h, and *DN*300 short pipes are used for other adjacent compartments. It is recommended to add an air shower room between the isolation suit take-off room and protective suit take-off room. This can effectively remove the virus absorbed by the suits of the medical staff.

3. Patient Entrances and Exits, and Other Auxiliary Rooms

A mechanical exhaust system is required at the patient entrances and exits, garbage rooms, and sewage channel, and the exhaust air volume can be based on the ventilation frequency of 12 times/h.

4. Restroom

It is recommended to locate outdoor restrooms downwind of the high-frequency wind direction in epidemic areas. According to the atmospheric diffusion model, it is recommended that the restroom be properly located in the place where there are no residential areas within a radius of 150 m.

2.4.4　Suggestions

I. Reservation of Ventilation Materials

All materials of the ventilation system need to be stored for disaster recovery.

II. Reserved Oxygen Pipelines

Oxygen pipelines need to be pre-laid for public buildings, which is suitable for converting into makeshift hospitals.

III. Principles for Ventilation and Air Conditioning Renovation

(1) Mechanical ventilation is adopted to ensure the orderly pressure gradient of the "three areas" (hygienic area, contaminated area and semi-contaminated area) and gradual reduction of static pressure. Each area is provided with an independent ventilation system.

(2) The exhaust fan/standby exhaust fan is arranged at the end of system and located outdoors. The air from the exhaust fan is discharged at high altitude after high-efficiency filtration.

(3) The entrance for medical staff adopts positive pressure control while the exit for medical staff, entrance/exit for patients, and material flow channel are subject to negative pressure control.

(4) The air supply system and exhaust system are interlocked, and the sequence of startup and stop is subject to the indoor static pressure requirements.

(5) The air outlet is arranged downwind of the dominant wind direction during the epidemic. The fresh air inlet and air outlet shall keep an appropriate distance to avoid mutual influence.

(6) The preferred indoor temperature is 14–28℃, and other temporary auxiliary facilities can be used when the indoor temperature cannot reach the preferred values.

(7) The ward and treatment area with all-air system should apply an all-fresh air operation mode.

(8) For the ward and treatment area

with cold/hot ends and fresh air system, the fresh air system should be renovated into the exhaust system for natural air supplement. Indoor negative pressure should be maintained. The exhaust system must be able to operate continuously.

（9）The condensate pipes of air conditioning in semi-contaminated and contaminated areas shall ensure reliable water tightness. Condensate water shall be collected and discharged to a designated point.

2.4.5 Summary

When the existing large space is transformed into makeshift hospitals for COVID-19 cases, the ward and treatment area's air conditioning and ventilation system incorporates an all-fresh air direct-current operation mode, which is simple and easy to implement. The exhaust air volume of the ward and treatment area shall be at least 90 m^3/h per person.

Presented reasonable ventilation methods for all areas can avoid cross infection and influence on surrounding environment, and also meet the extremely short time requirement for transformation.

Countermeasures against the outbreak of respiratory infectious diseases and rules to be followed for ventilation and air conditioning transformation are provided.

II

技术创新篇
TECHNOLOGICAL INNOVATION

3

技术创新

3.1 BIM 数字孪生

3.1.1 项目背景

"数字孪生"这个概念的基本雏形是 2003 年美国密歇根大学教授迈克尔·格里夫斯（Michael Grieves）提出来的，他称其为"物理产品的数字化表达"[23]。我们现在理解数字孪生，认为它是现实世界的实体在虚拟数字世界的镜像，这个镜像不仅能虚拟再现现实实体，还能模拟其在现实环境中的行为。建筑行业的数字孪生技术主要基于 BIM 技术，根据近几年国内的发展现状，BIM技术的应用关注点从单一的建筑物几何特性越来越向建筑全生命周期的设计、建造、运维

一体化管理倾斜，但是，由于使用 BIM 技术会造成一次性投资的增加，国内 BIM 技术的应用普遍比较单一，无法真正实现 BIM 建造的数字孪生模型在建筑全生命周期内的价值。

2020 年初暴发的新冠肺炎疫情，势必会促进下一步国内医疗基础设施建设，作为政府投资的公共服务类项目，应该更加关注建筑在全生命周期内的使用效率、成本和能源消耗。通过抗疫应急医院——雷神山医院的 BIM 技术应用，笔者对医疗建筑的 BIM 技术应用提出了思考，以期为即将到来的医疗项目建设大潮提供一些建议。

雷神山医院（图 3.1）位于武汉市江夏区，是一个专为收治新冠肺炎重症、危重症患者建造的抗疫应急医院，建设用地面积约 22 万平方米，总建筑面积约 8 万平方米，可提供床位约 1500 张，容纳医护人员约 2300 名。项目根据用地情况主要分为东区隔离医疗区和西区医护生活区，并配备有相关运维用房，均为 1 层临时建筑。

3.1.2 雷神山医院 BIM 技术应用

雷神山医院设计建造的重点和难点主要有三个：一是要能快速建成并投入使用；二是要防止对环境造成污染；三是要避免医护人员感染。

图 3.1 雷神山医院规划总平面图及 BIM 模型鸟瞰图

医院采用模块化设计，呈现独特的鱼骨状布局，每根"鱼刺"都是独立的医疗单元，都是隔离医疗区（图 3.2、图 3.3）。根据雷神山医院项目的特点，送排风系统的主要管线均在室外敷设（图 3.4），那么一般传统的 BIM 应用点，例如管线综合、净高分析等，在雷神山医院项目中已经不再是关注焦点，于是，项目中 BIM 技术的应用就围绕上述三个项目重点和难点展开。

1. 基于 BIM 的数字化建造

雷神山医院要求十余天建成使用，建设工期紧是整个项目面临的主要问题。而建筑骨架是施工的第一道工序，所以结构专业的设计和施工速度直接影响整个项目的建设速度。为了解决以上问题，雷神山医院的隔离医疗区全部采用轻型模块化钢结构组合房屋体系，医技区根据对开间和净高的要求，采用钢框架结构体系。

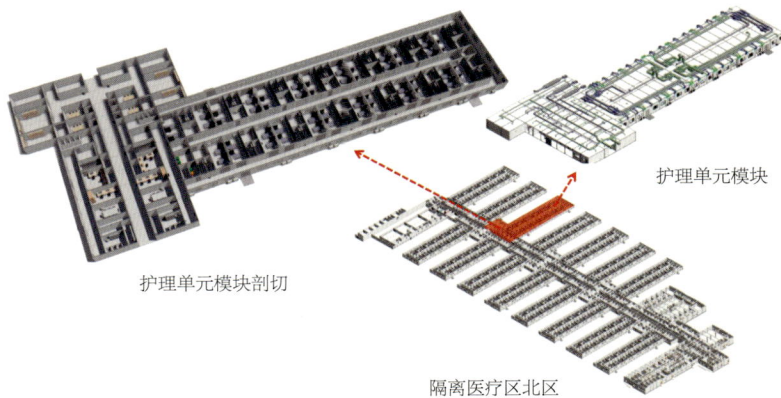

护理单元模块剖切

护理单元模块

隔离医疗区北区

图 3.2 隔离医疗区的 BIM 模型图

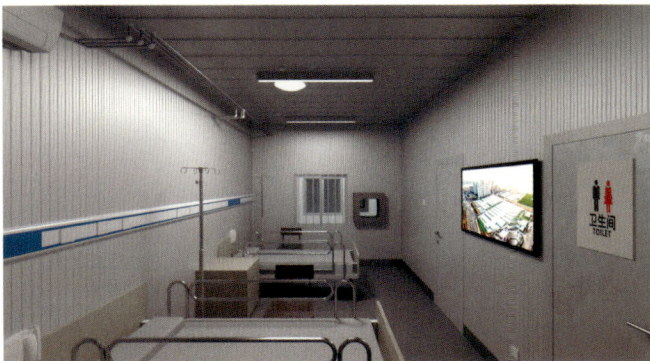

图 3.3 负压病房的室内 BIM 模型图

图 3.4 隔离医疗区实景

数字化应急医院设计及建造技术

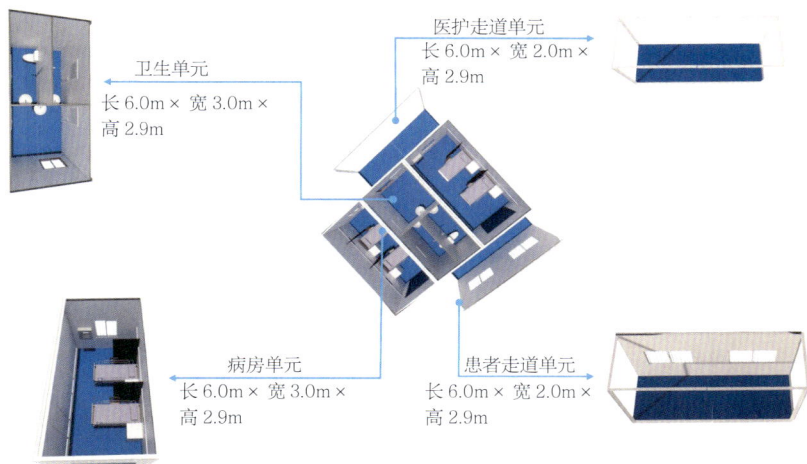

图 3.5　隔离医疗区模块化结构解析图

1个病区由4个功能模块组成（图3.5），利用基于BIM的数字化建造技术，将建筑和结构构件、机电设备在数字模型中进行集成和归类，直接指导工厂制作。同时对现场施工工序进行数字化模拟（图3.6），寻找最佳拼装方案，并根据功能和拼装顺序对模块进行数字编号，现场像堆积木一样进行施工建设，极大缩短了项目建设工期。

2. 室外风环境模拟分析

雷神山医院的建设选址非常严格，项目周围没有居民区，所有污水、雨水通过有组织地收集处理、消毒后，排入市政管道，是绝对安全的。病区的排风也经过高效过滤后进行排放，但仍希望医院排放的气体能迅速在空气中扩散稀释。于是项目利用BIM模型进行风环境的分析（图3.7）。

从分析结果来看，建筑物周围未形成死角或者漩涡区，场地通风情况良好，有利于场地内排放的气体迅速被稀释和扩散。

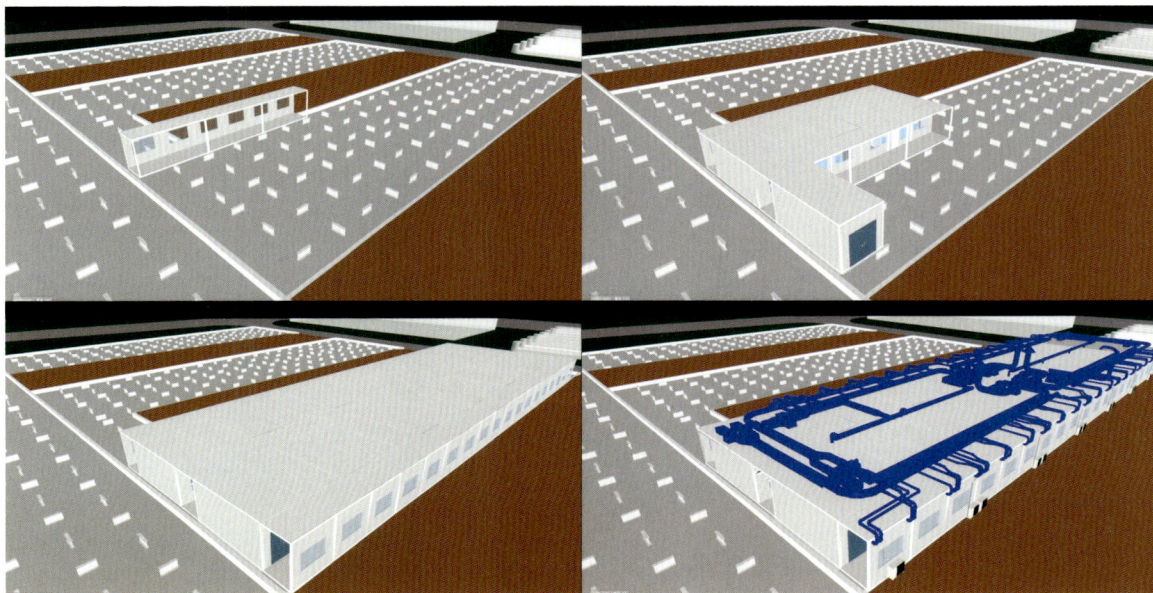

图 3.6　隔离医疗区施工数字化模拟

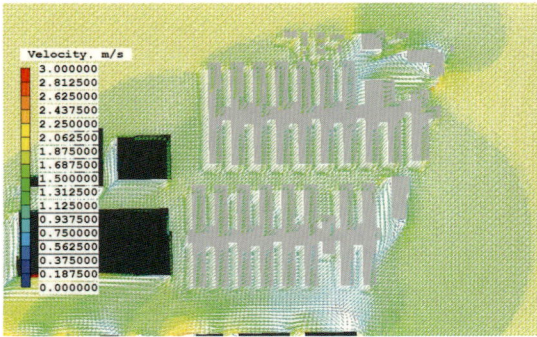

图 3.7　项目场地自然通风风速矢量图

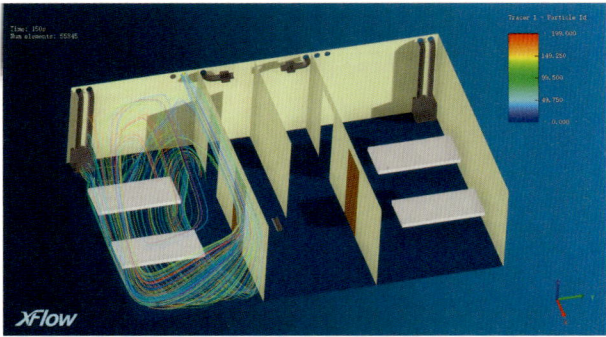

图 3.8　负压病房气流流动轨迹

3. 负压病房气流组织模拟

负压病房是医护人员感染的重灾区，对负压病房气流组织的分析，旨在辅助设计，为医护人员的安全提供保障。

通过分析，在如图 3.8 所示的送排风布局下，病房内形成了 U 形通风环境，气流从送风管流出，碰到对侧墙壁后改变方向，最后流经 2 位患者后到达下部回流区，经排风口过滤后排出，这种通风环境能有效改善病房内污染空气的浓度，降低医护人员感染的风险。

4. 雷神山医院 BIM 技术应用总结

雷神山医院项目作为特殊时期的特殊产物，BIM 技术应用主要围绕项目设计建造的重点和难点展开，对项目的快速建成并投入使用起到了一定的促进作用，但在 BIM 技术的延伸与拓展方面，仍存在一定的局限性，在未来医疗建筑的建设、使用中，应通过数字孪生医院拓展应用，在建筑全生命周期中强调数字信息管理，使其发挥更大的效能。

3.1.3　数字孪生医院的应用思考

在民用建筑中，医院是最复杂的建筑类型之一：功能分区复杂、医疗工艺设计复杂、使用空间要求多变、能耗非常大、医疗设备管理维护复杂等。据统计，一座大型医院的各科设备管线达 40 余种。鉴于这些特点，一座医院的建设成本在其全生命周期成本中只占一小部分，

而使用阶段的能源消耗、设备维护、系统管理等成本将占大部分，所以像医院这样的公共事业项目，应在建设阶段就考虑其在全生命周期内的消耗和产出。利用 BIM 技术建立医院的数字孪生模型，并基于此模型进一步搭建运维平台，前期虽有一定投入，后期其价值产出却能极大地节约成本，提升医院运营效率。

1. 医疗建筑的 BIM 标准化设计

医疗建筑设计机构根据医疗建筑功能的特点，用 BIM 技术对医疗建筑中的功能模块进行拆分和标准化设计，将设计转化为成熟的产品，有利于提升设计质量和效率，如病房区的病房模块、缓冲区模块、医技区的手术室模块、ICU 区模块、CT 室模块等。将大模块拆分成重复使用率高的小模块，有利于建筑、结构、机电、装修、设备等全专业进行精细化、参数化建模，添加材质、尺寸、设备参数等构件信息，方便同类型医疗建筑的设计提取和复用（图 3.9 ～ 图 3.11）。针对特殊时期的应急项目，利用成熟的 BIM 模块化产品进行 BIM 正向设计，可以大大提升设计质量和速度，所见即所得。进一步更可延伸至装配式建筑的工厂加工、现场拼装，解决应急项目工期紧张的问题。

2. 基于 BIM 的数字化建造

相较于以汽车制造业为代表的先进制造业，目前建筑行业的数字化应用程度远远不够，随着技术的进步和发展，利用数字化建造技术，像造

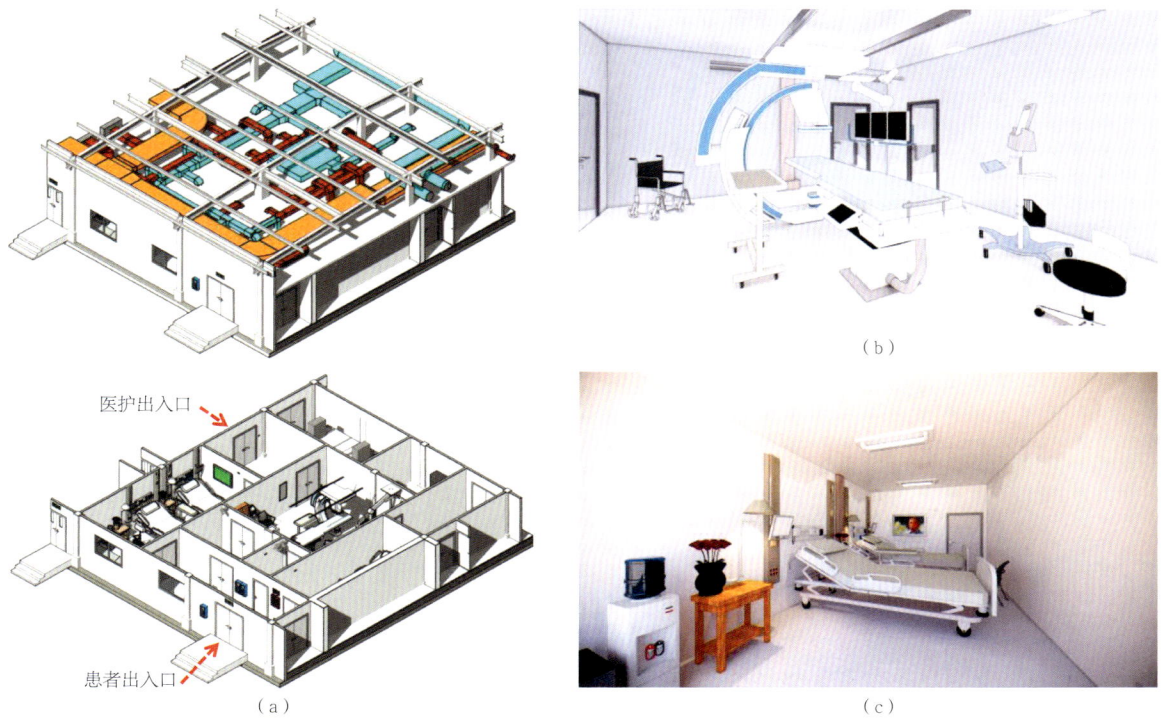

图 3.9 负压手术区 BIM 模型图
（a）负压手术区模型；（b）手术室模型渲染图；（c）复苏室模型渲染图

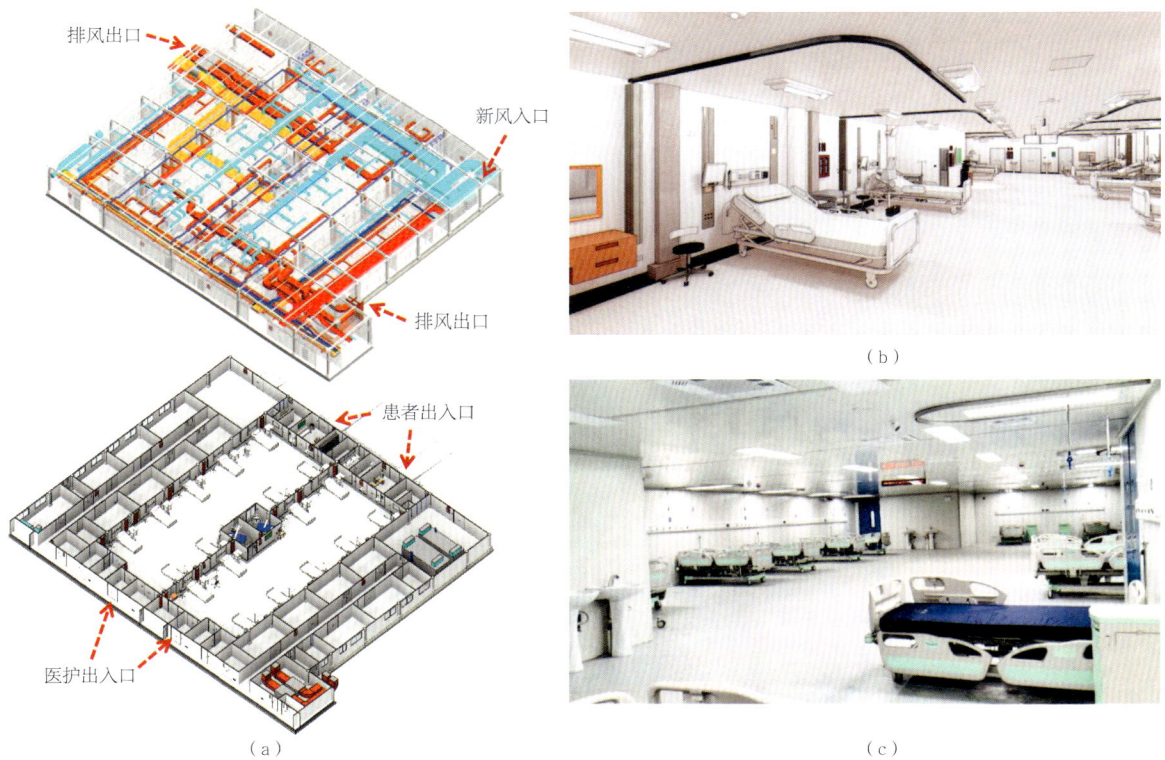

图 3.10 负压 ICU 区 BIM 模型图与实景照片
（a）负压 ICU 区模型；（b）负压 ICU 区模型渲染图；（c）负压 ICU 区实景照片

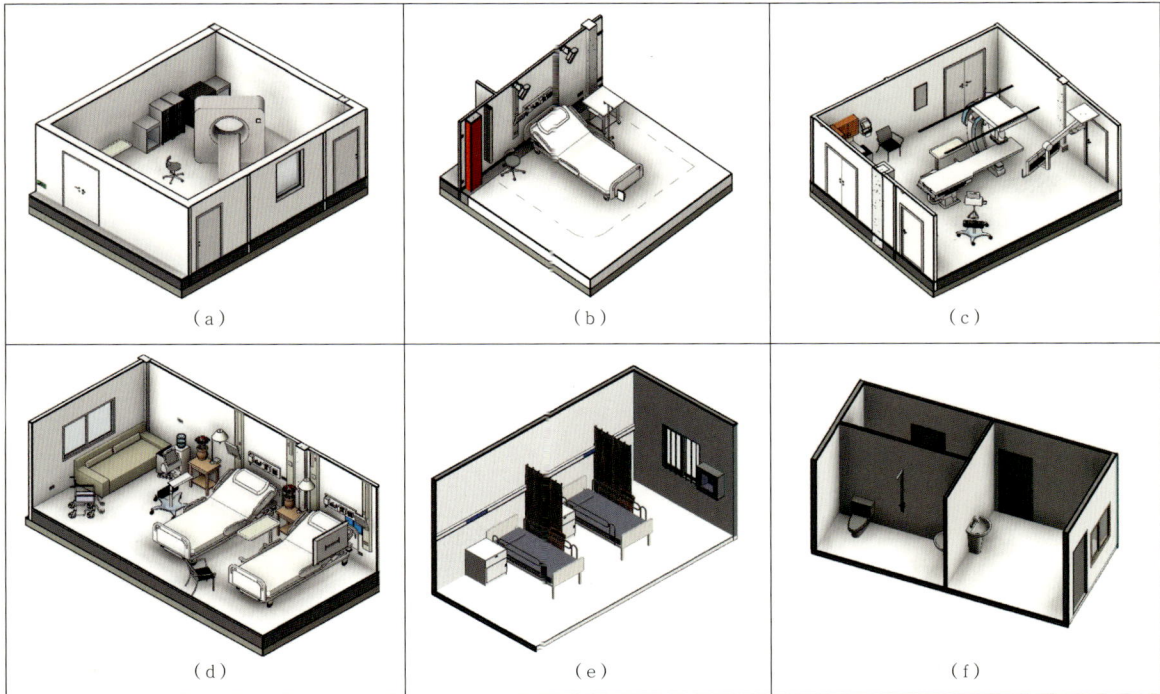

图 3.11 医疗建筑 BIM 标准化模块

（a）CT 室模块；（b）ICU 模块；（c）手术室模块；（d）复苏室模块；（e）病房模块；（f）卫生间 + 缓冲室模块

汽车一样造房子，是我们共同的努力方向。通过建立 BIM 模型，除在设计阶段的一系列应用外，在施工阶段还可以进行施工工序模拟、现场数字化放样、工厂数字化加工、物料追踪、工程进度质量管理、实时计量支付等。特别是对于医疗建筑中标准化程度较高的病房，可以利用 BIM 模型集成更多专业构件，进行数字化工厂加工和生产，采用装配式技术进行现场拼装，可提升施工效率和质量，并能最大限度地减少物料损失，降低环境污染，这也是未来的发展方向。

3. BIM 协同平台的应用

医院在建造阶段，数字孪生模型的建立依靠 BIM 技术，除了当前应用较多的管线综合、建筑物性能分析、施工模拟、工程造价等功能，更应强调数据模型的三维协同。这种协同不只是设计阶段的各专业协同，更是基于可视化技术所有参建方、后期运营方、使用方集合的大协同（图 3.12）。利用基于云平台的网络协同平台，展示各方成果，收集各方意见，各方信息互通、实时协同，同时可以进行进度、质量、造价的协同管理。在设计建造过程中更多考虑智慧运营的需求（图 3.13）。

4. 数字孪生医院在运维中的应用

数字孪生的应用维度有两个，一个是强调物理特性的几何模型，另一个是强调数字应用的管理模型，两者结合，才能使数字孪生技术发挥出最大的效能。一般运维阶段的数字应用依靠大数据、物联网等技术进行建筑物空间、资产、设备、能源等管理。针对医疗建筑，主要有以下几个方面值得我们重点关注。

（1）大型医院内部的就诊导航和行程定制

大型综合医院规模大，内部流线复杂，传统二维标识系统导向效能较弱，可通过数字孪生医院的三维空间属性，在手机端实现就诊导航和行程定制，提升患者就诊体验。此外，还可规划紧急抢救流线、感染疫情控制流线、无障碍流线等。

（2）楼宇设施设备全过程精准运营维护

医院的设施设备是医院核心价值最高的部

医护走道

负压病房

图 3.12　雷神山医院 BIM 模型可视化漫游

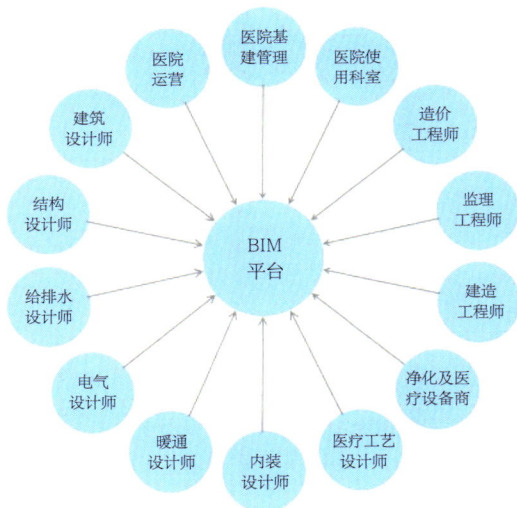

图 3.13　医院项目 BIM 平台信息传递示意图

分，高效率地运营维护不仅能节省能源及整体运行费用，更能确保医院运行安全。传统的以 BA 系统（楼宇自控系统）为基础的智能化平台体系，由于缺乏各系统组件所在空间形态的准确定位及运行全矢量参数的支撑，维护可及性远远不能满足医院运行的实际需求。数字孪生医院提供的建筑设施的数字化、空间化、可视化模型，结合 BA 系统技术可实时监控建筑内部能源使用

情况，实现设备系统及时准确的故障派修及预防性保养维护[24]。

（3）医疗设备的数字化管理

通过 BIM 模型，医院可将 CT、MR、检查设备、检验设备等医疗重资产设备信息更准确地与医院空间关联起来[25]，并在数字孪生医院中储存设备参数信息，实现设备虚拟实体与文本参数的整合统一，保证设备资产管理的准确性与及时性，利于医院进行资产管理。

（4）大数据和人工智能

结合人工智能，通过对医院运营的全仿真，可统计就诊人群数量、就诊需求、就诊时间等医疗空间使用数据，按照需求变化适时调整医院空间结构，并可在数字孪生医院中进行预改造，结合大数据和 VR 系统进行预判断与预决策。同时，建立的数字模型还可模拟消防疏散、公共突发事件等，提升医院安全性能。

（5）环境监测平台

环境监测平台利用 BIM 模型、仿真软件、互联网技术共同构建，具有响应快、跨地域、不间断、大规模、适应性强等特点（图 3.14）。中

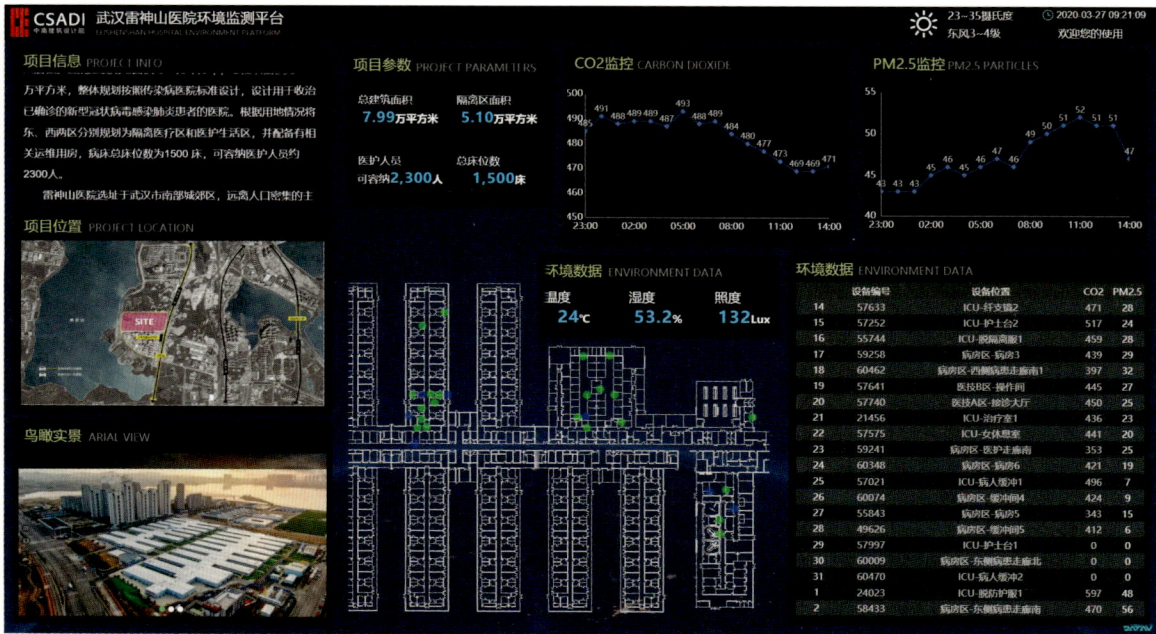

图 3.14　环境监测平台界面

南院与清华大学合作进行雷神山医院和多个方舱医院的室内环境监测，通过识别算法优化通风系统和净化设备运行，优化医院日常运行。

还要说明的是，数字孪生医院不仅可在新建医院使用，在既有医院实施基于竣工模型的数字化管理，对既有医院的管理提质增效有立竿见影的效果，未来还将有广阔的市场和应用空间。

3.1.4　结语

2015 年住房和城乡建设部发布的《关于推进建筑信息模型应用的指导意见》中明确提出，到 2020 年，由国有资金投资为主的大中型项目，集成应用 BIM 技术的项目比率达到 90%。目前我们离这个目标还很远，医院作为有着巨大市场前景的公共事业项目，应更有理由在 BIM 技术的应用上抢先一步。

BIM 技术的应用是雷神山医院设计、建造的重要支撑，基于此，我们将进一步开展利用 BIM 技术建立数字孪生医院的研究工作，为即将到来的智慧化时代做好准备，也同时为随机而来的突发事件做好准备。

3.2　装配式建筑的应用

3.2.1　技术策划

为应对新冠肺炎疫情，急需新建一座武汉雷神山医院救治患者，新增床位 1500 张，总建筑面积约 8 万平方米，必须在十余天内完成。中南院临危受命，火速组建设计团队，进行技术策划，制定了采用装配式钢结构建筑的设计建造方案。

装配式建筑是当前国家推行绿色建筑的主要抓手。中南院较早开始装配式建筑设计和科研工作，先后主编、参编了多部国家、省、市装配式建筑的相关标准，是中国工程建设标准化协会发布的《模块化装配整体式建筑设计规程》（T/CECS575—2019）等技术标准的主要编制单位。此次武汉雷神山医院的设计正是在这样的技术储备基础上进行的。

武汉雷神山医院采用装配式钢结构设计建造。通过技术策划，在装配式建筑的结构系统、外围护系统、设备管线系统和内装系统上，按照工业化快速建造的要求进行模块一体化集成设计，利用 BIM 信息化管理技术，在各个参建

　数字化应急医院设计及建造技术

方的共同努力下，顺利完成了这样一个全球关注的特殊工程项目的建设。

中南院设计团队特别重视技术策划工作，全面分析该项目的特点（表 3.1）、难点及关键点，通过细致分析各种相关影响因素，寻找合适的项目技术策略。

通过分析上述特点，剖析项目实施的难点，寻找相关对策和措施，见表 3.2。

按照这个技术策划方案，各专业各部门立即开始了紧张的设计施工工作。

3.2.2 建筑设计

按照前面的技术策划，需要围绕整体项目进行功能分区设计。建筑功能区包括隔离医疗区、医护生活区、保障功能区三大分区。其中，隔离医疗区按照模块装配式建筑设计理念进行整合归并，按相同模块进行组合拼装（图 3.15）。整个隔离医疗区呈鱼骨状单层分布。全部单元可以分解为两种基本单元（单元平面尺寸为 3.0m×6.0m 及 2.0m×6.0m）进行排列组合，模块单元的高度为 2.9m。模块由主体结构、楼板、墙板、吊顶、设备管线、内装部品组合而成，模块构成功能集成的三维空间体。医技区采用普通装配式钢结构建筑。医护生活区包括宿舍区、办公区、餐饮区及清洁用品库等。宿舍区有 10 栋宿舍楼，均为标准化设计的 1 层或 2 层

项目特点分析   表 3.1

序号	项目特点	描述
1	极短工期	十余天（正常设计施工周期应在 180 天以上）
2	超大规模	约 8 万平方米，床位约 1500 张，医务人员约 2300 名
3	建筑功能复杂	超强感染性传染病医院，其人流、物流流线复杂
4	场地条件好	场地条件比较好，适合天然基础
5	项目定位高	举全国之力，用中国速度高质量建造

项目难点、对策及措施分析   表 3.2

序号	难点	对策	措施
1	传染病医院建筑功能复杂	将复杂的功能区标准化、模块化分解，各个模块细分设计	对相同或者相近功能区进行标准化设计，方便施工
2	多专业配合复杂	多专业采用一体化集成设计	利用 BIM 平台共享设计数据
3	极短工期完成设计与施工装修	部分项目单层标准化布局同步设计与施工，争分夺秒	对单层标准模块进行装配式建筑分解，充分考虑方便施工，无须电梯安装
4	模块建筑加工制作量大，巨量采购	细分不同功能模块，分类同步加工生产	总包单位从全国各地寻找合适的生产工厂，同步制作采购
5	现场施工人员众多	将整体项目单层施工，模块化分解，化整为零	全场单层分块同步施工，加大施工作业面
6	多工序交叉施工	采用模块一体化集成设计，减少主体结构、设备安装及装饰等中间环节	采用模块装配式集成房屋，各项建筑功能在工厂一次成型，现场整体安装，局部调整

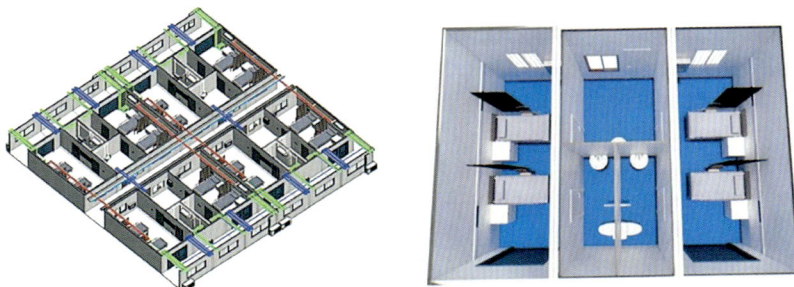

图 3.15 隔离医疗区病房模块单元组合

装配式钢结构建筑，建筑高度为7.5m。宿舍区的大部分建筑位于原来军运会的场馆内，采用标准化的轻钢装配式建筑，建造快捷，技术成熟。

3.2.3 结构系统

模块装配式钢结构建筑是技术高度集成的装配式建筑，工业化程度远高于其他类型的装配式建筑。天津大学、广州大学及中南院等单位分别主编了模块装配式建筑的国家或者行业标准，国内众多的钢结构制作企业研究开发了模块箱式房屋的加工制作技术。正是在这些技术研究及应用的基础上，迅速开展了武汉雷神山医院的模块装配式建筑设计应用。

在武汉雷神山医院工程设计中，针对隔离医疗区、医护生活区不同的建筑功能和空间特点，分别采用了模块装配式钢结构和普通装配式钢结构两种形式。

1. 模块装配式钢结构

隔离医疗区的病区护理单元具有典型的标准模块化的特征，故采用模块装配式钢结构建筑体系（图3.16）。

这种模块箱式单元采用钢结构骨架和彩钢复合板墙体，结构整体性强，承载力高，抗风、抗震性能好，安全适用。模块箱式单元可单独使用，也能根据使用需求进行多元化组合，自由拼接，通过水平及竖直方向的不同组合和拓展形成宽敞的使用空间。武汉雷神山医院项目仅使用单层模块组合，使用的总模块数有3190余个。

2. 普通装配式钢结构

（1）隔离医疗区的医技区建筑层高为4.3n，平面柱网不统一且局部跨度达到18m，选择采用普通装配式钢结构建筑，采用钢框架结构体系，如图3.17、图3.18所示。

（2）医护生活区建筑为2层，根据建筑平面和空间要求，采用轻钢活动板房体系。其主体结构单元为轻钢结构框架，为了保证结构的

图3.16 病区护理单元模块箱式单元现场安装

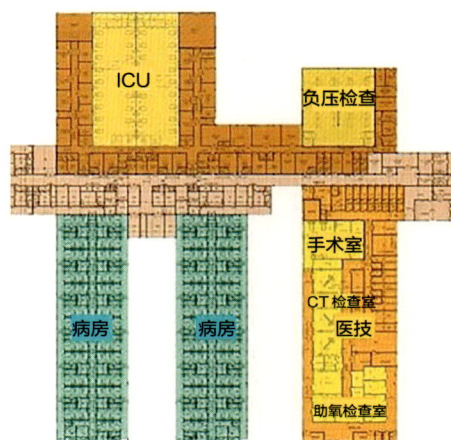

图3.17 隔离医疗区医技区局部平面

图3.18 隔离医疗区医技区现场安装实景图

抗侧性能，框架间设有交叉拉索，增强了结构的抗侧刚度和安全性能。

楼面支承体系采用轻型钢桁架或H型钢梁，其上铺设结构板，屋盖及墙体围护采用夹芯彩钢板。这种活动板房技术体系成熟，采用标准化、模数化生产，安装和拆卸非常方便及快速，用

数字化应急医院设计及建造技术

于临时应急建筑具有明显优势，能满足建筑的各项功能性需求。

3.2.4 外围护体系

武汉雷神山医院采用了两种类型的装配式建筑。模块装配式钢结构建筑的外围护体系是与模块箱式房屋建筑、结构、设备和装修等功能一体化集成设计的，并由工厂一体化制作。

武汉雷神山医院的外围护体系除要遵循一般外围护系统的标准要求外，还有严格的气密性要求，单个模块箱式房屋在工厂按标准制作，其外围护墙体的气密性能够满足要求，但在箱体拼接成更大的开间时，要跨越箱体外墙板接缝。箱体间的缝隙用泡沫胶填充密实，墙板上的开孔与拼接的缝隙，先用密封胶封闭，再覆盖铝箔，确保气密性满足要求。

普通装配式钢结构建筑外围护结构为岩棉复合板，型材尺寸为 1820mm（长）×950mm（宽）×100mm（厚）。岩棉复合板选用 2 层高品质的彩色涂层钢板或其他精密压型的金属板材作面板，成型后的板材具有优良的防火、保温、环保等特性。

3.2.5 设备管线系统

装配式建筑设备管线系统的设计要求体现在标准化设计、模块化集成设计及管线分离设计等方面。

1. 标准化设计

武汉雷神山医院项目机电设备及管线采用标准化设计，设备管线、部件、接口尺寸实现标准化、系列化，确保采购便利，部品部件通用性强、互换性强，安装快捷。

（1）给排水设计和安装标准化。病区与非病区的室内外给水、排水（包括排水通气）管必须独立设置（图 3.19）。隔离医疗区卫生间和隔离间设计成一个标准的卫生单元（图 3.20），医

图 3.19　隔离医疗区室外 BIM 模型图

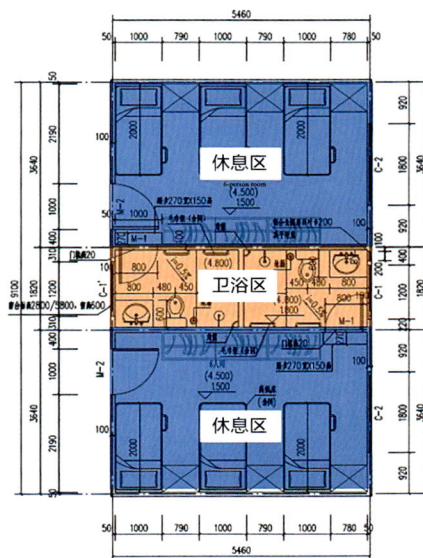

图 3.20　医护生活区典型单元

护宿舍内卫生间采用淋浴房，专家楼采用整体卫浴等多种形式的集成式卫生间。对墙面（板）、吊顶、洁具设备及管线等进行集成设计，在模块箱体屋面四个角分别设 *DN*50 雨水排水管将雨水排至箱体地面以下的架空层。

（2）暖通设计中 80% 以上的通风空调风管选用标准化成品 PE 管道，采用的管道规格根据系统的划分计算确定，干管不做变径。现场仅须与支管简单电熔连接，安装便捷。

负压隔离病房通风设备采用成品箱式离心风机，设备出厂前已将风机、电机及减震器集成至一个箱体内，现场可快速与风管连接。新风系统过滤采用的粗效、中效及高效过滤器也在生产工厂内集成为高效过滤单元，无须现场逐一拼装各部件。

2. 模块化集成设计

借助 BIM 技术，机电设备及管线采用模块化集成设计，主要体现在以下几个方面。

（1）给水蓄水箱采用装配式水箱，模块化设计，全部采用断流水箱供水。给水排水设备、管线与主体结构分离，方便维修更换，且在维修更换时不影响主体结构。

（2）标准负压隔离病房采用模块化设计。暖通专业根据负压隔离病房的要求，结合送风排风系统气流组织的技术特点，进行了 CFD 模拟。根据模拟结果，精确定位室内送风口与高效排风口的位置（见图3.8）。隔离病房内采用设备带集成氧气、负压吸引、压缩医用气体终端以及开关、电源插座、呼叫等终端（图3.21、图3.22）。

（3）负压手术室自成单元模块。手术室的净化空调系统、通风空调系统风口、医用气体设备结合手术室其他设施、设备，采用机电一体化集成设计，手术室自成模块。其中，过滤及通风部件采用装配式构造，高效过滤送风单元集成至吊顶内（图3.23），空调回风口及过滤器集成至装配式墙面内，空调采用成品单元式空调机组，手术室实现整体模块现场拼装。

（4）采用箱式变电站及箱式静音柴油发电机组。相比室内机房而言，由工厂提前加工预制，更方便施工和快速就位，检修也方便安全。

3. 管线分离设计

武汉雷神山医院项目采用管线分离设计，给水排水、通风空调和电气设备管道与主体结构分离，采用干法施工。其中，医院负压隔离病房通风空调干管设置在箱式房屋面，医技用房干管设置在公共区吊顶内，风机及空调设备设置在屋面或清洁区机房内。病房内医用气体管道与墙体分离，安装在设备带内，如图3.24所示。

3.2.6 内装系统

装配式建筑的内装系统设计遵循建筑、装

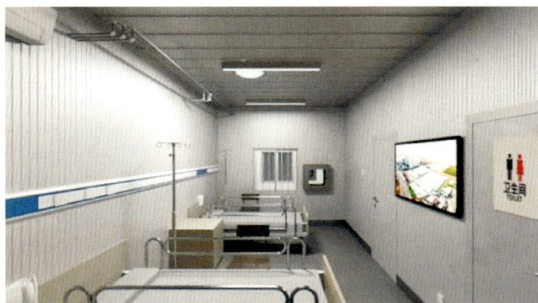

图3.21 隔离病房 BIM 设计效果图

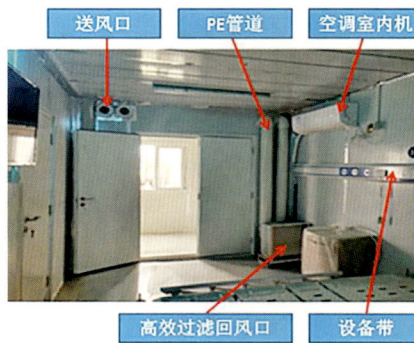

图3.22 隔离病房设备安装实景图

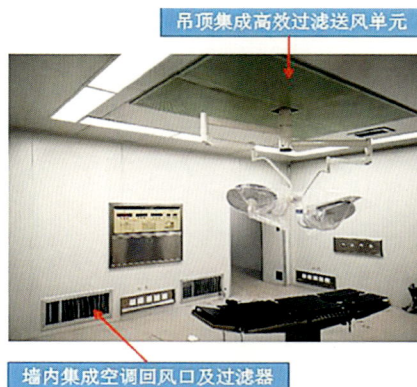

图3.23 手术室高效过滤送风单元及回风口集成

来源：曹晓庆，张银安，刘华斌，等. 雷神山医院通风空调设计[J/OL].（2020-02-27）[2021-12-15]. https://kuaibao.qq.com/s/20200227A05Y1T00?refer=spider.

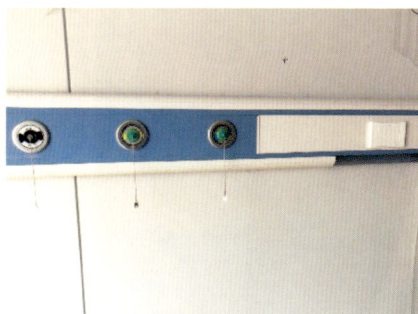

图3.24 隔离病房设备带终端集成图

饰、部品一体化设计原则，采用工业化生产的内装部品，实现集成化的成套供应。通过优化参数、公差配合和接口技术，提高部品构件的互换性和通用性。内装部品具有可变性和适应性，便于施工安装、使用维护和改造等。雷神山医院采用的模块装配式箱式房屋，实现从建筑、结构、装饰、部品一体化设计到工厂一体化建造，现场直接安装，局部适当调整即可交付使用，对医院的及时建成起到关键作用。医技区和生活区的普通装配式钢结构建筑的围护板材也都是工业化生产的保温装饰一体板，安装连接方便。医院各个区域的卫生间按工业化部品安装，淋浴房大都采用整体淋浴房。大多数机电设备实现集成化的成套供应。这些部品、构件在工地安装方便，为在极短工期内完成约8万平方米的装配式建筑建造创造了条件。

3.2.7 信息化管理与制作安装

利用数字化模拟建造的方式进行装配式建筑设计和施工，可提升装配式建筑的设计和生产效率。采用BIM技术将建筑和结构构件、机电设备进行集成和归类，可直接指导工厂制作。同时利用BIM技术的可模拟性对现场施工进行模拟，寻找最佳拼装施工方案，可大大提高现场的拼装效率，加快施工速度。武汉雷神山医院设计施工时间太短，来不及进行BIM全模型正向设计，但在医技区装配式钢结构的主体结构设计时，设计与钢结构制作单位紧密配合，直接将设计模型与工厂生产的数据对接，施工图即为加工详图，直接导入车间生产，节省了大量宝贵的时间。

在时间紧、任务重的紧急条件下，设计、施工、制作与安装高度协调与配合。设计遵循标准化、模数化与集成化原则，尽可能地利用成熟工业化产品体系。施工过程配合中，采用诸多变通方法，如采用贝雷梁替换原架空层结构，并针对柱脚安装等问题进行了处理。在此过程

中，也积极采用先进的信息化管理方法（BIM技术）解决本项目中千军万马交叉作业的矛盾。

3.2.8 结语

（1）通过创新的技术策划，采用模块装配式钢结构建筑模式，利用高度的技术集成，实现建筑、结构、设备及内装一体化建造，优质、快捷、高效。武汉雷神山医院中多种创新技术的成功应用充分说明模块装配式建筑体系强大的应用价值和未来巨大的发展空间。

（2）结构与基础形式的多种灵活选择，为整个工程的顺利实施赢得了宝贵的时间。

（3）装配式建筑设备管线系统的标准化、模块化和管线分离设计，在武汉雷神山医院获得了成功的应用，效果良好。

（4）装配式建筑设计、施工、制作、安装高度协调与配合的一体化建造模式，正是中国建筑行业转型与升级的发展方向。设计与施工联合的工程总承包（EPC）模式能使工程项目实施全过程高效运作，武汉雷神山医院的设计与施工一体化，充分显示出这种模式的强大生命力。

3.3 数值仿真应用

3.3.1 项目概况

为了满足临时医院的防疫要求，雷神山医院的通风系统需要进行针对性的设计。在医院设计时采用了负压病房空气系统，能够有效防止病毒扩散至医院内部，从而避免病毒扩散引起交叉感染。在负压隔离病房中，合理的通风系统能够有效地改善病房内空气质量，有助于病人的康复，并且最大限度地保护医护人员。此外，雷神山医院还设计了专门的排风系统，从高空向室外排出病房内受污染的空气，进而降低医院周围其他区域受到污染的风险。

传统的设计方法无法从雷神山医院病房内

含病毒空气浓度的角度对不同设计方案进行定量比选，也无法评估雷神山排出含病毒空气对外界的影响。不管是病房内通风系统的建设，医院初期的选址，还是医院运行时废气排放的二次污染防控，都急需一种能够有效模拟病毒空气扩散机制的分析方法来提供更加定量、具体的参考和依据。此外，为满足雷神山医院这种设计和建设周期较短的临时医院，所选用的数字化分析方法不仅需要足够的计算精度，还需要较高的计算效率。

CFD 技术能够定量分析含病毒气体扩散的轨迹及含病毒空气在空间上的浓度分布，为临时医院的设计和选址提供参考。在众多 CFD 分析软件中，基于格子玻尔兹曼方法（Lattice Boltzmann Method，LBM）[26] 的 XFlow 软件兼具宏观流体连续模型和微观分析动力学方法的优点，具有高效的计算效率和良好的并行性[27]。

3.3.2 分析流程概述

进行 CFD 分析一般包括三个步骤：①建立分析对象的几何模型，包括负压隔离病房、雷神山医院及医院周边的建筑物；②根据实际情况选择相应的边界条件和计算假定，并在 CFD 软件中生成计算模型；③进行计算结果后处理及可视化。考虑到临时医院项目的设计和建造周期，有必要采用高效的三维建模工具。Rhino 软件具有高效的参数化建模功能[28]，能够高效地建立多个建筑物模型，并导入支持大多数三维格式的 XFlow 中接力进行 CFD 模型处理及流体力学计算，最后还可以利用 XFlow 自带的后处理功能对计算结果进行可视化分析。

1. 负压隔离病房污染空气扩散

（1）负压隔离病房简介

雷神山医院是由标准模块装配而成的拼接式建筑，病房区域的负压隔离病房单元采用规格统一的集装箱，如图 3.25 所示。为了研究雷神

山医院病房内通风系统的送风和排风口布局对污染空气扩散的影响，分析考虑了四种不同的送排风布局，如图 3.26 所示。这四种通风系统根据送风口和排风口的不同，分为：①同侧送风、下排风；②对角送风、下排风；③居中送风、两侧对角下排风；④居中送风、排风。采用 Rihno 软件建立如图 3.26 所示的标准病房模型之后，便可以在此基础上根据病房的不同设计方案，基于参数化建模功能快速得到相应的建筑物模型，为通风系统的多方案比选提供便利。

（2）计算域网格划分和时间步长

采用 XFlow 进行 CFD 分析时，计算域的格子划分形式直接影响计算的精度和效率。在进行负压隔离病房污染空气扩散分析时，在送风口、排风口、污染源区域采用精细化格子，分别为 31.25mm、6.25mm、7.8125mm，而在其他区域采用 125mm 的较粗尺度格子，如图 3.27 所示。计算模型的总格子数为 14 万，这样既能保证所关心区域的计算精度，也能有效提升分析的计算效率。

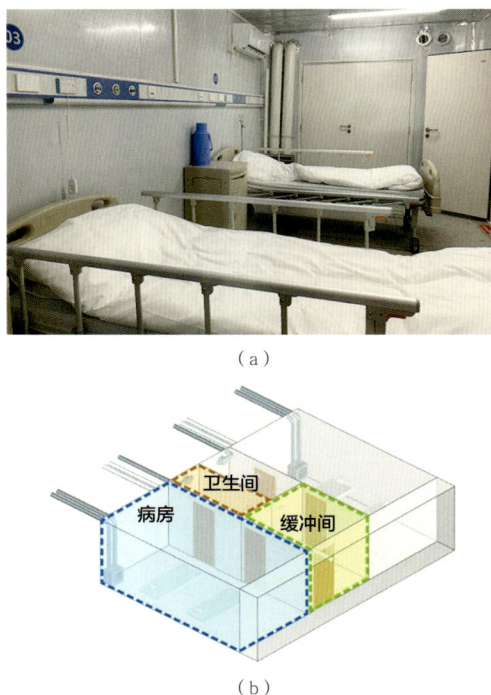

（a）

（b）

图 3.25 负压隔离病房实景图和模型图
（a）实景图；（b）模型图

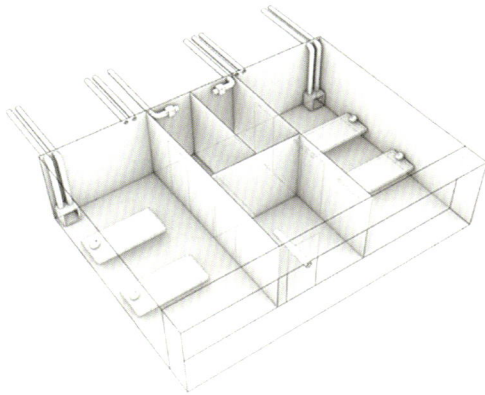

图 3.26　四种通风方案对比

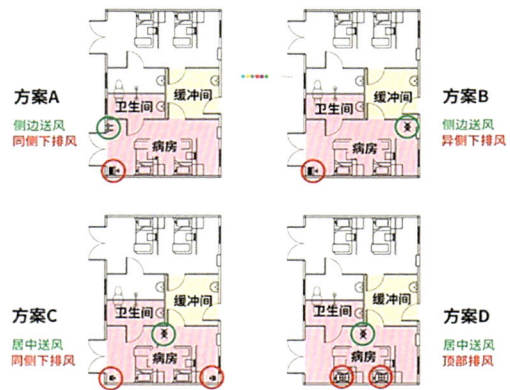

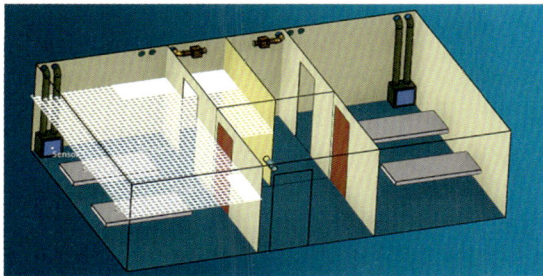

图 3.27　负压隔离病房网格划分

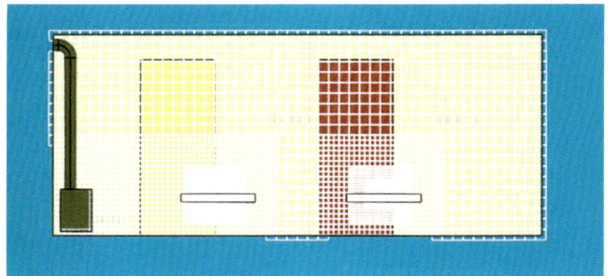

XFlow 为瞬态分析软件，选取足够长的计算时间才能够获取污染空气充分扩散后的稳定状态。针对负压病房污染空气扩散分析，选取 300s 作为计算时长，时间步长为 0.0001s。

（3）病房内边界条件

针对负压病房污染空气扩散分析，需要确定的边界条件包括通风口的压力边界条件、风速边界条件及释放源的浓度。病房采用直流式通风系统，通过定风量阀控制送排风支路的风量恒定。房间设计总排风量为 700m³/h，总送风量为 500m³/h。病房和缓冲间送风口的速度边界条件分别为 3.93m/s 和 1.57m/s。病房设计相对压力为 −15Pa，卫生间设计相对压力为 −20Pa，排风口采用压力边界。假定病人为含病毒空气的释放源，由于释放源与病人之间的差异相关，所以这里采用相对浓度进行分析。因此，假定病人释放的含病毒空气浓度为 1，房间初始浓度场为 0，后文介绍的浓度均为相对浓度。

2. 医院外部污染空气扩散

（1）雷神山医院及周边建筑

为了研究雷神山医院排出的污染空气对周围群众的影响，分别选取了吹向住宅区域的东北风和吹向医院生活区的东南风工况下的污染空气扩散分析。按照实际情况分别建立雷神山医院和周围建筑的分析模型，其实景图和模型图如图 3.28 所示。

（2）计算域网格划分和时间步长

为了避免计算域边界对气流扩散分析结果产生不利影响，需要保证足够大的计算域的尺寸，本次分析建立了尺寸为 4000m× 10000m×300m 的城市级数值风洞，如图 3.29 所示。在进行雷神山医院外部污染空气扩散分析时，位于 ICU 病房出口处的格子采用 125mm，位于普通病房出口处的格子采用 250mm，远离医院区域的格子最大尺寸在 16m，计算模型的总格子数为 519 万。针对雷神山医院外部污染空气扩散分析时，选取

图 3.28 雷神山医院及其周边建筑实景图和模型图

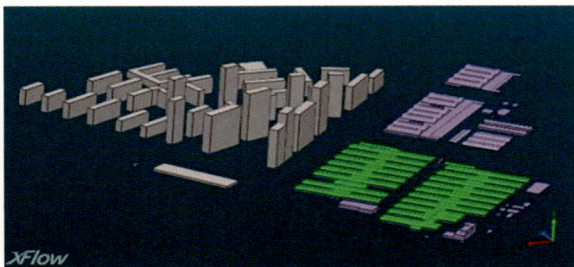

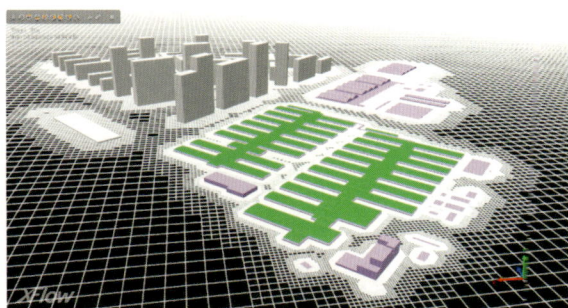

图 3.29 雷神山医院分析模型网格划分

图 3.30 医院各病房污染空气排量

图 3.31 监控的关键区域示意图

50min 作为计算时长,时间步长为 0.067s。

（3）医院外部排放边界条件

针对雷神山医院外部污染空气扩散分析时,ICU 和普通病房的污染空气排量如图 3.30 所示,其中假定通风系统的过滤器能够将排出的污染空气稀释 1000 倍。室外污染空气需要考虑来流风的特性,分别选取对医院周围住宅区域较为不利的东北风工况及对医院生活区较为不利的东风工况,风速均取武汉地区冬季最常见的风速 3m/s。

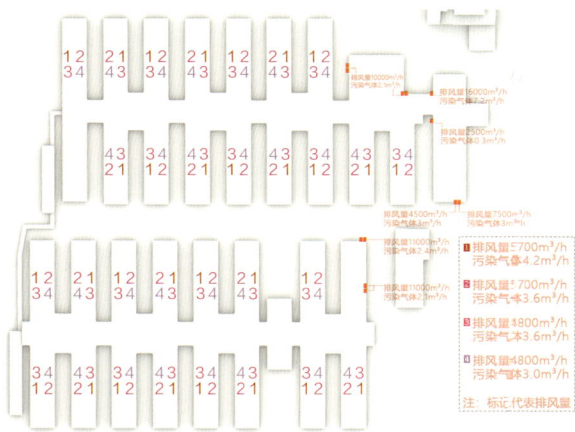

3.3.3 病房内污染空气扩散情况

1. 通风系统方案对比

XFlow 软件能够输出气体流动轨迹、流动速度及污染空气在空间上的浓度分布,通过对这些结果的分析能够对不同通风系统的效率进行定量对比,从而使得设计人员能够更加全面地对设计方案进行优化。在 CFD 建模时,需要将分析对象的关键区域设置为监控点,以便于在后处理时能够提取相关的数据。针对负压隔离病房的分析,选取了排风口及医护人员停留的区域和病床区域作为关键区域进行监控,如图 3.31 所示。

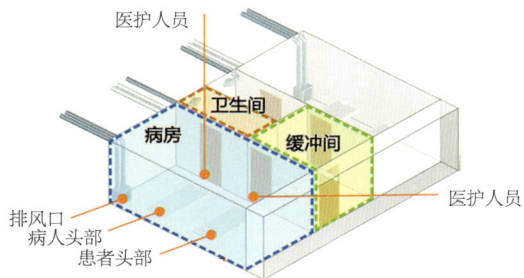

（1）气流流动轨迹

图 3.32 给出了四种不同的通风方案的气流流动轨迹。A 方案中送风口和排风口布置在同一面墙壁,大部分气流由送风口传递至病房另一侧的墙壁后,经过患者头部区域,之后到达排风口,其余部分气流继续在房间内循环。B 方案送风口和排风口在房间对角布置,气流多在

送风口附近循环。C方案为居中送风、两侧排风，气流大多经过床底向房间两侧循环，而房间中部的气流较少。D方案为居中送风和排风，气流经过病房两侧的墙壁反射之后通向排风口，气流同样大多经过床底。从气流流动的区域来看，A和D两个方案中气流在房间内的循环较充分。

（2）污染空气相对浓度

下面选取三个分别代表排风口、患者面部及医生站立的不同高度的横截面对病房内的污染空气相对浓度场进行分析，比较这四种不同的通风系统。

图3.33给出了排风口高度的横截面污染空气相对浓度云图，其中A方案排风口处的污染空气相对浓度较高，病房内其他位置污染空气相对浓度较低，这表明污染气体随气流到达排风口，A方案中的污染气体可以有效排出。其余三个方案中排风口高度的截面污染空气相对浓度普遍高于A方案。

图3.34给出了患者面部高度截面污染空气相对浓度云图。其中靠近观察窗的患者上方到排风口污染空气相对浓度相对较高，污染气体可以随气流有效扩散至排风口位置，病房内其他位置污染空气相对浓度相对较低。其余方案患者上方截面污染空气相对浓度均高于A方案。

图3.35给出了医护人员站立高度的截面污染空气相对浓度云图。其中A方案送风口下方三角区域污染空气相对浓度相对较高，污染气体随着气流循环后，病房内其他区域的污染空气相对浓度较低，相对来说，绿色区域污染空气相对浓度较低，是较安全的区域。其余方案患者上方截面污染空气相对浓度均高于A方案。

提取如图3.30所示的各监控点的污染空气相对浓度值列于表格3.3，其中A方案除了监控点2的污染空气相对浓度高于D方案，其余监控点病房污染气体平均浓度均低于其他方案。总的来说，采用A方案的通风系统，病房内的

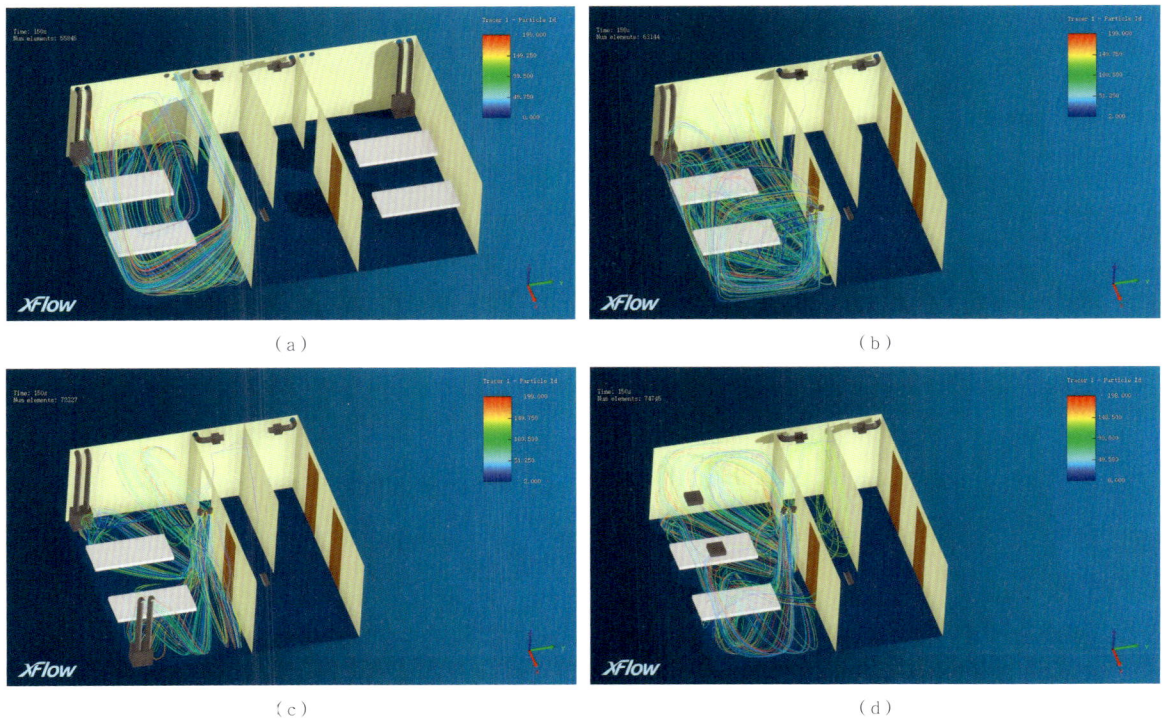

图3.32　四种方案的气流流动轨迹对比
（a）A方案；（b）B方案；（c）C方案；（d）D方案

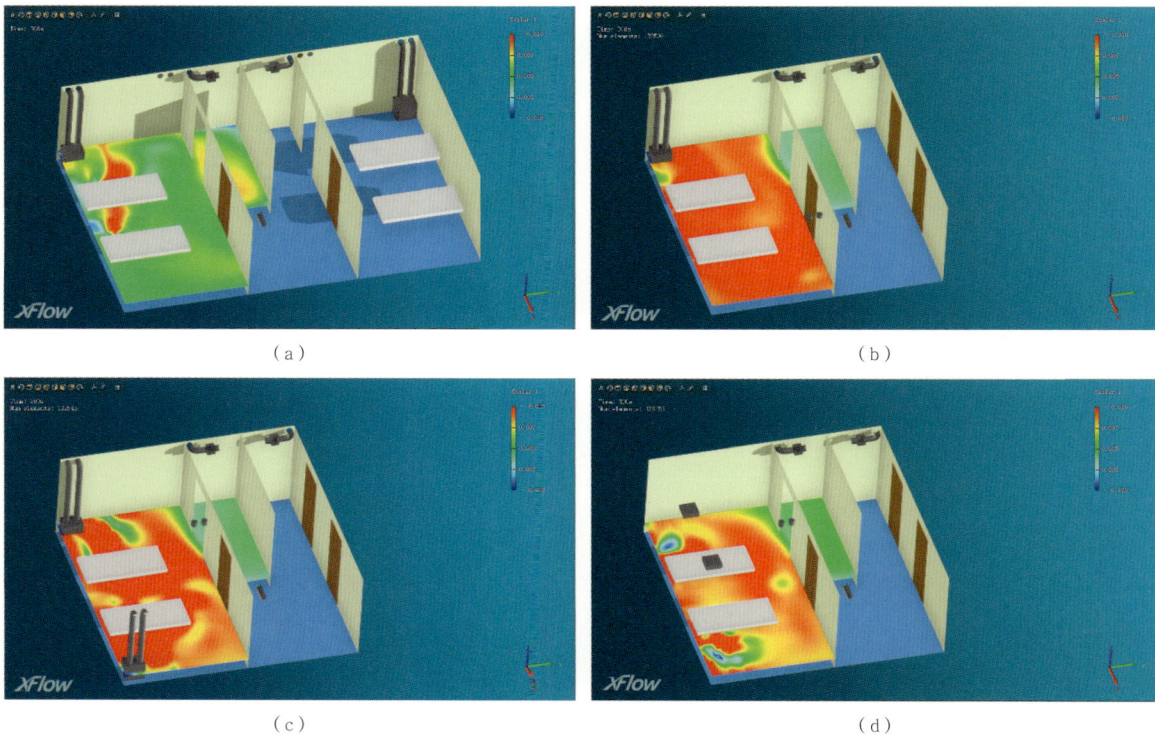

图 3.33　四种方案的污染空气相对浓度对比（排风口高度）
（a）A方案；（b）B方案；（c）C方案；（d）D方案

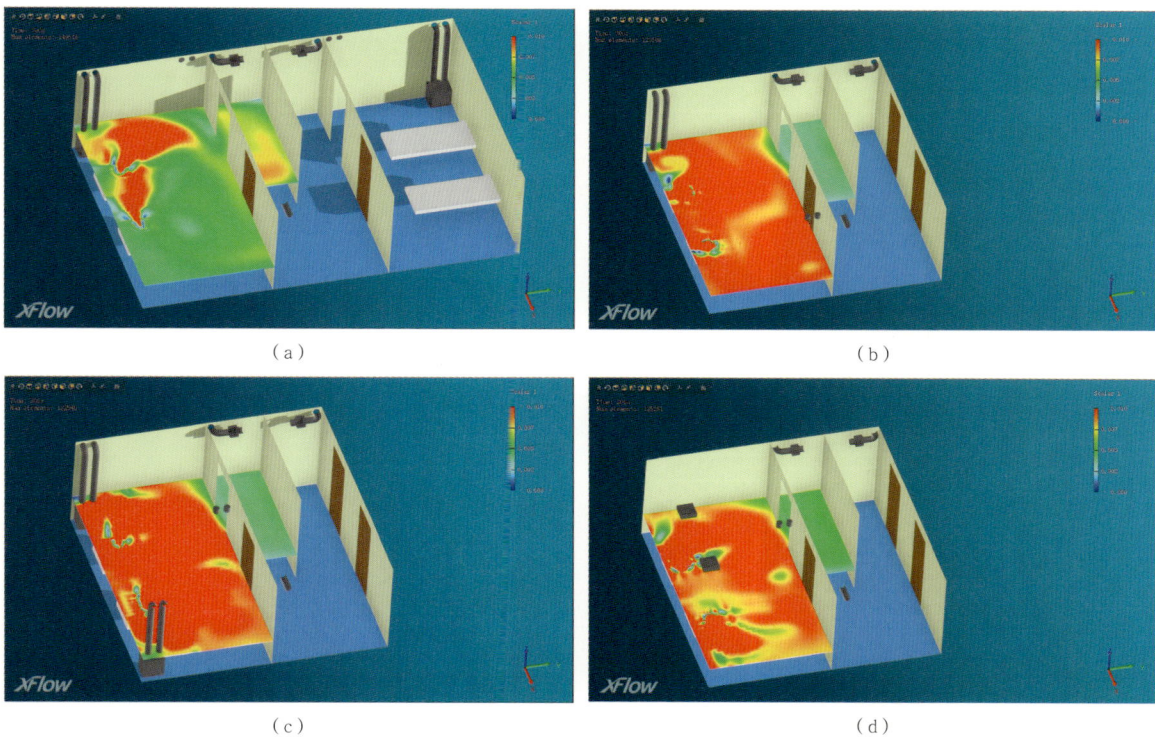

图 3.34　四种方案的污染空气相对浓度对比（患者面部高度）
（a）A方案；（b）B方案；（c）C方案；（d）D方案

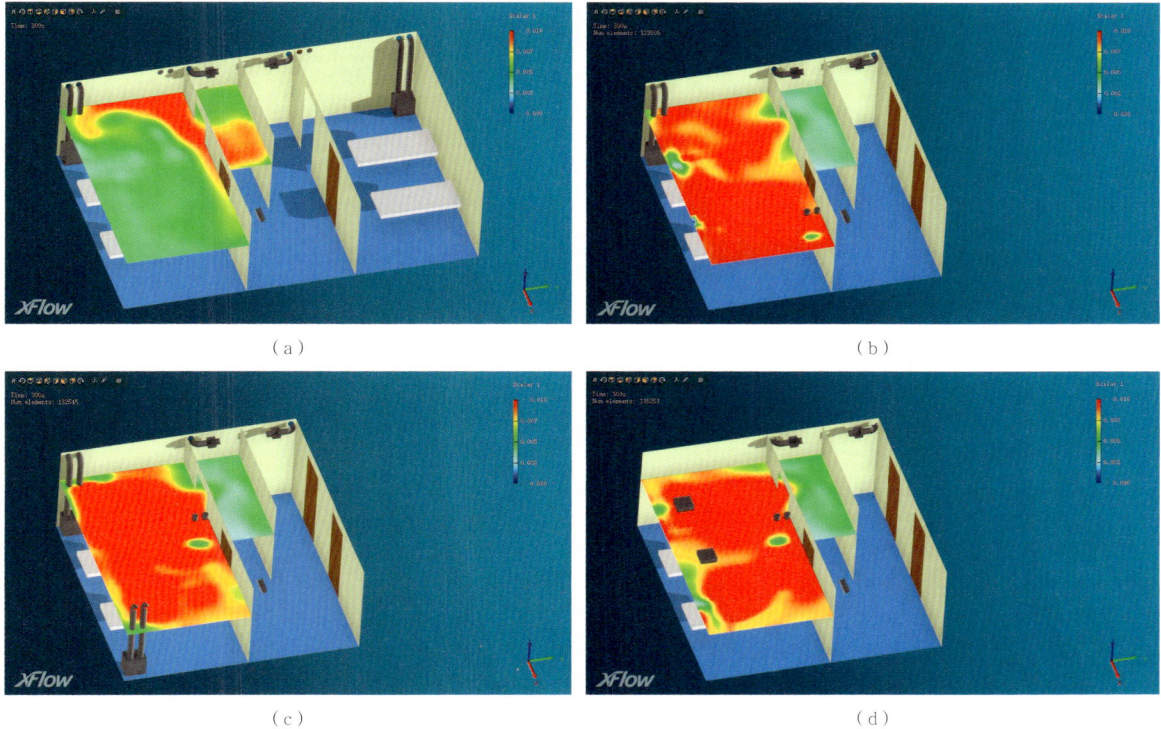

（a）

（b）

（c）

（d）

图 3.35　四种方案的浓度对比（医生站立高度）

（a）A 方案；（b）B 方案；（c）C 方案；（d）D 方案

四种方案的污染空气相对浓度对比 表 3.3

通风方案	排风口点 1	病人 1 点 2	病人 2 点 3	医生 1 点 4	医生 2 点 5	病房平均
A 方案	0.0069	0.0052	0.0077	0.0044	0.0046	0.006
B 方案	0.0082	0.0691	0.1661	0.0097	0.0149	0.009
C 方案	0.0087	0.0068	0.0128	0.0097	0.0138	0.011
D 方案	0.0086	0.0024	0.1622	0.0112	0.0231	0.009

污染空气相对浓度较低。

2. 卫生间和缓冲间的影响

雷神山医院负压隔离病房的功能分区如图 3.22（b）所示。在整个护理单元内部，病房与卫生间和缓冲间之间都分别通过一扇门进行通行。由于病房内卫生间的门和缓冲间的门有时会处于开启状态，这对病房内的污染气体的浓度会有一定的影响。选取以下四种工况进行分析：①卫生间和缓冲间全关闭；②仅卫生间开启；③仅缓冲间开启；④卫生间和缓冲间门全开启。

表 3.4 列出了病房内污染空气相对浓度趋于稳定之后的四种工况下，各监测点及房间内的污染气体平均浓度。图 3.36 给出了四种工况下排风口处污染空气相对浓度随时间变化的曲线，结合表 3.4 的房间污染气体平均浓度变化曲线可以看出，排风口处的污染空气相对浓度与房间的污染气体平均浓度相关。

通过对比"门全关闭"和"卫生间开门"工况的浓度可以发现，卫生间门开启后，排风口的污染空气相对浓度降低而其余监测点的污染空气相对浓度普遍升高，房间的污染气体平均浓度也升高。这是由于开启卫生间门促使室内气流湍流强度上升，这种气流紊乱导致排风口的效率降低，从而引起了病房内污染空气相

四种工况下污染空气相对浓度对比

表 3.4

计算工况	排风口点 1	病人 1 点 2	病人 2 点 3	医生 1 点 4	医生 2 点 5	病房平均
门全关闭	0.0118	0.0036	0.0086	0.0033	0.0032	0.0038
卫生间门开	0.0069	0.0051	0.0077	0.0044	0.0046	0.0058
缓冲间门开	0.014	0.003	0.007	0.0031	0.003	0.0025
门全开启	0.012	0.003	0.0072	0.0036	0.0029	0.0035

对浓度的升高。通过对比"缓冲间开门"和"门全开启"的工况，可以进一步印证开启卫生间会导致病房内污染空气相对浓度升高这一结论。

通过对比"门全关闭"和"缓冲间开门"工况的污染空气相对浓度可以发现，缓冲间门开启后，排风口的污染空气相对浓度升高而其余监测点的污染空气相对浓度普遍降低，房间的污染气体平均浓度也降低。这是由于开启缓冲间门之后，缓冲间与房间内部的压差导致气流的流动轨迹发生了改变。这种气流循环路径的改变结合 A 方案排风口的布置使得病房内的通风效率得到了一定程度的提升。通过对比"缓冲间门开启"和"门全开启"的工况，可以进一步印证开启缓冲间门会降低病房内污染气体平均浓度这一结论。

总的来说，从污染空气相对浓度分析的角度，卫生间的门处于开启状态不利于病房内污染气体的排出，而缓冲间门开启时，由于压差的存在，对污染气体的排出是有利的。

3.3.4 医院外部污染空气扩散情况

在冬季东北风和东风工况下，人呼吸高度处的污染空气相对浓度如图 3.37 所示，针对东北风工况，重点关注医院周边住宅区 6 栋建筑物的污染空气相对浓度情况，而针对东风工况，重点关注医院生活区的污染空气相对浓度情况，如图 3.38 所示。

冬季东北风向影响的居民区域主要包括办公区与住宅区中间三角分布的 6 座建筑及附近的街道，其中迎风面建筑底部人呼吸高度处（Z=1.5m）最大浓度是 ICU 出口的 1.6%，假定出风口污染空气稀释倍数为 1000 倍，则迎风面建筑底部人呼吸高度处（Z=1.5m）稀释倍数为 62500 倍。

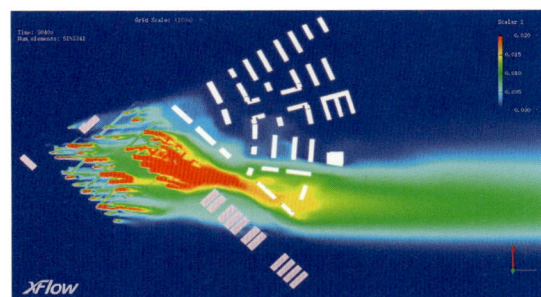

（a）

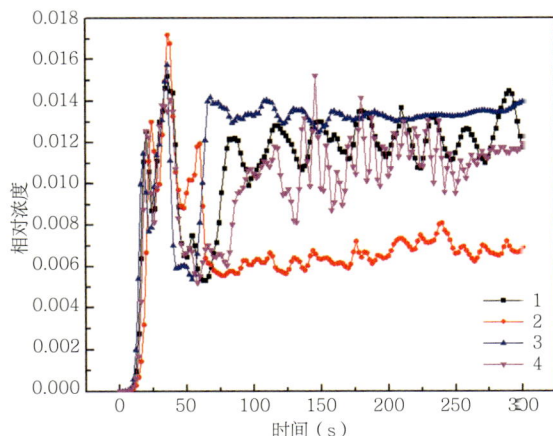

（b）

图 3.37 雷神山医院外部 1.5m 高度处污染空气相对浓度云图
（a）东北风工况；（b）东风工况

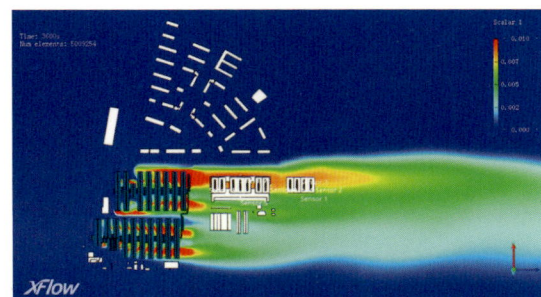

图 3.36 四种工况下排风口的污染空气相对浓度对比

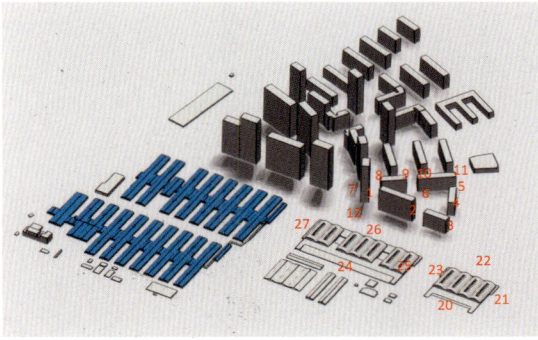

图 3.38 住宅区和医院生活区主要监控点

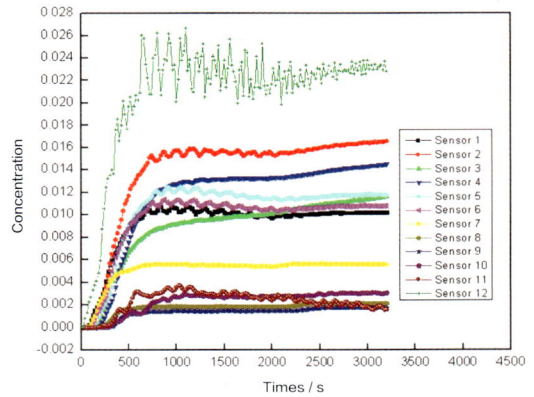

（a）

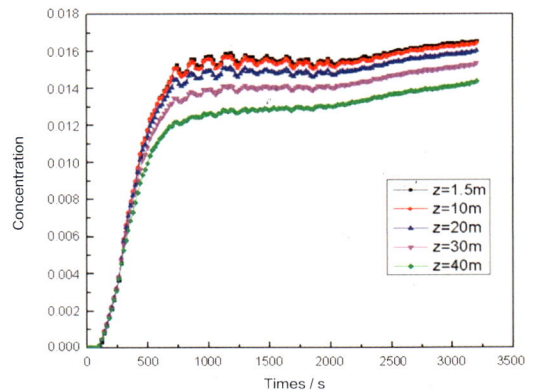

（b）

图 3.39 住宅区污染空气相对浓度曲线
（a）各监控点浓度随时间变化曲线；（b）浓度随高度变化曲线

附近街道 $Z=1.5m$ 高度处浓度是ICU出口的2.3%（稀释倍数为43478倍），如图3.39所示。

东风工况下，主要受影响的区域为雷神山医院的食堂及宿舍区域，食堂迎风面的污染空气稀释倍数最小为27900倍，宿舍一层最小稀释倍数为16900倍，宿舍区域5m高度处最小稀释倍数为27900倍，如表3.5所示。

3.3.5 结论

本部分内容主要介绍了以无网格流体分析软件 XFlow 为载体的CFD技术在雷神山通风系统设计中的应用。利用领先的流体分析软件XFlow模拟了负压病房的气流组织和污染空气扩散，通过对比多个通风系统布置方案，找到了污染空气扩散最小的方案，最大限度地保护医护人员和防止交叉感染。

（1）四种方案中A方案的通风效率较高，房间内的污染空气相对浓度基本低于其他方案。

（2）开启卫生间门不利于病房内气体的排出，建议门应保持关闭状态。

（3）当缓冲间门与病房内的门开启时，缓冲间与病房之间的压差有利于病房内污染气体的排出。

此外，建立城市级大尺度的流场模型，模拟了医院病房外废气排放对周边环境的影响，为医院整体规划设计及医护人员、患者、公众的健康安全提供了依据。

（1）东北风工况作用下，住宅区域的含病毒气体被稀释到了43478倍以上。

（2）东风工况作用下，医院生活区含病毒气体被稀释到了16900倍以上。

根据相关研究[29]，SARS病毒被稀释到了10000倍以上时，就不再具有传染性。因此，可以认为雷神山医院对其他区域的影响较小，不会影响到公众的安全。

医院生活区监控点污染空气相对浓度对比   表3.5

监控点	20	21	22	23	24	25	26	27
东北风	2.11×10^8	2.01×10^8	5.02×10^7	6.67×10^7	1.90×10^6	9.45×10^6	1.64×10^5	1.32×10^5
东风	3.14×10^5	4.05×10^5	2.98×10^5	2.79×10^5	3.02×10^5	3.58×10^5	2.10×10^5	1.69×10^5

3

The Innovation in Design Technology

3.1　BIM Digital Twin

3.1.1　Project Background

Digital twin is a virtual replica of physical entities, namely the representation of real entities in the digital virtual world. Digital twinning in the construction industry is mainly achieved by BIM technology. Leishenshan Hospital is an emergency temporary hospital established in 2020 to fight against COVID-19. BIM technology technically supports its rapid construction and safe operation.

The major three challenges in the design and construction of Leishenshan Hospital are that it needs to be completed and put into operation quickly, to prevent environmental contamination, and to avoid infection of medical staff.

The hospital adopts modular design and presents a unique "fishbone-shaped" layout. Each "fishbone" is an independent medical unit and an isolation ward and treatment area (Figure 3.1). According to the characteristics of this project, the main pipelines of the air supply and exhaust-system are all laid outdoors. The traditional BIM applications are no longer the focus in this project, such as pipeline integration and clearance analysis. Therefore, the application of BIM technology in this project focuses on the above three difficulties.

3.1.2　Application of BIM Technology in Leishenshan Hospital

I. BIM-based Digital Prefabrication

Leishenshan Hospital is required to be built and put into service in about 10 days. The negative pressure ward area applies light steel modular structure. The medical

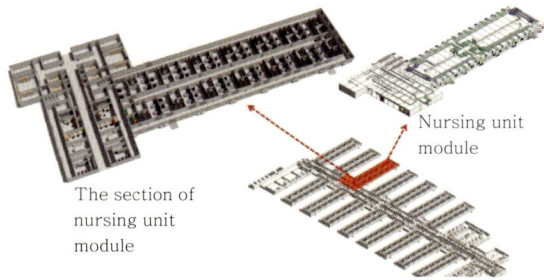

The section of nursing unit module

Nursing unit module

Figure 3.1   Section Rendering for BIM Model of a Ward and Treatment Area (Top View)

test and imaging area adopts steel-framed-structure due to the requirements on the width and the clearance.

A ward and treatment area includes four function modules (Figure 3.2) . BIM technology is used to integrate and classify buildings, structural components and electromechanical equipment in the digital model, which can directly provide guidance for factory production ; it is also used to digitally simulate the on-site construction procedures (Figure 3.3) to find the best assembly scheme. As the assembly can be realized just as stacking toy blocks, the construction period will be greatly shortened.

II. Simulation Analysis of Outdoor Wind Environment

Although the contaminated air of the ward and treatment area is exhausted after efficient filtration, it is expected that they can diffuse and dilute rapidly in the air. Therefore, this project adopts BIM model for wind environment analysis (Figure 3.4) .

The analysis results show that no dead space or vortex area around the building and the sound ventilation are conducive to the rapid dilution and diffusion of discharged waste gases.

III. Air Distribution Simulation in the Negative Pressure Ward

The analysis of air distribution in the negative pressure ward (Figure 3.5) aims to find the best scheme for air supply and exhaust, thereby facilitating the rapid circulation and exhaust of contaminated gases and reducing the risk of cross-infection in the ward.

The analysis results show that the air supply and exhaust layout shown in the figure

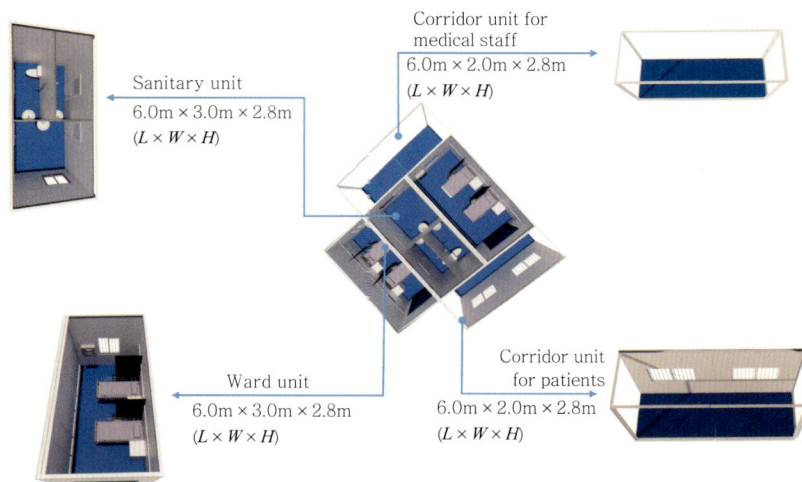

Sanitary unit
6.0m × 3.0m × 2.8m
($L \times W \times H$)

Corridor unit for medical staff
6.0m × 2.0m × 2.8m
($L \times W \times H$)

Ward unit
6.0m × 3.0m × 2.8m
($L \times W \times H$)

Corridor unit for patients
6.0m × 2.0m × 2.8m
($L \times W \times H$)

Figure 3.2   Analytical Diagram of Modular Structure in Negative Pressure Ward Area

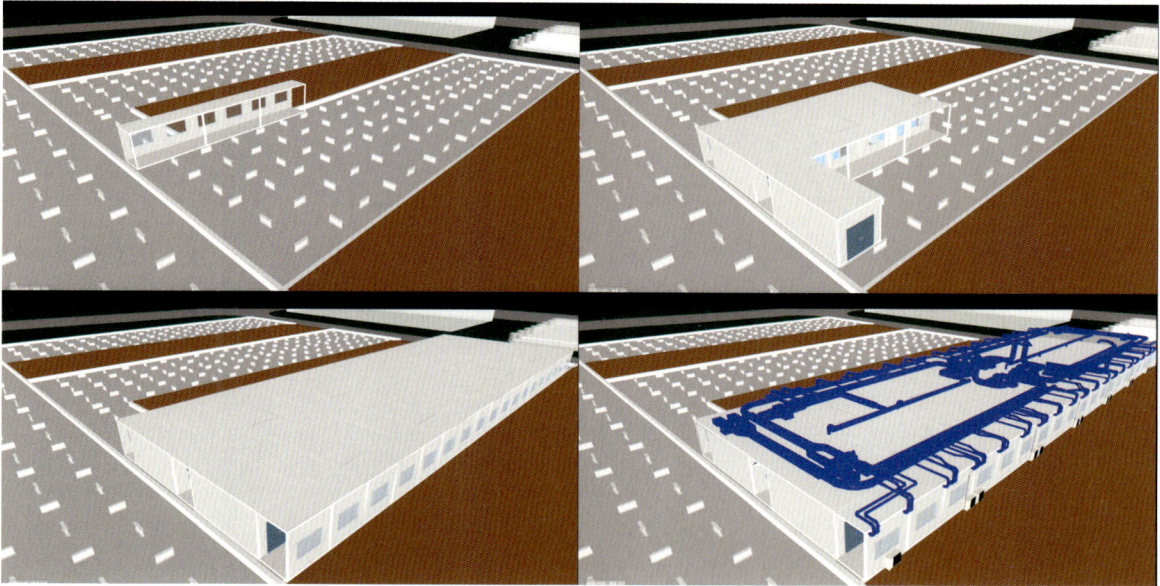

Figure 3.3   Digital Simulation of Construction of Negative Pressure Ward Area

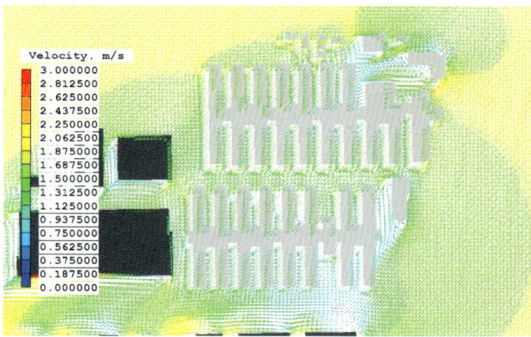

Figure 3.4   Vector Diagram of Wind Speed under Natural Ventilation at the Project Site

Figure 3.5   Airflow Trajectories in the Negative Pressure Ward

formed a "U" type ventilation environment. The airflow comes out from the air supply pipe, changes direction after encountering the opposite wall, then flows through the two patients, and finally reaches the lower recirculation zone. It is exhausted after being filtered at the air exhaust point. This ventilation environment can effectively decrease the concentration of contaminated air in the ward and reduce the risk of infection to medical staff.

3.1.3   Thinking on Application of Digital Twin Hospital

Hospital is the most complex functional public buildings and its construction cost is only a small portion throughout the project. The energy consumption, equipment maintenance and system management will account for most during the operation period. Therefore, the public utility projects, such as hospital, should consider making use of digital twin technology in construction to improve hospital efficiency in the whole life cycle.

I. Standard BIM Design of Healthcare Facilities

Based on functional features of healthcare facilities, related design institutions should apply the BIM technology to divide them into such modules as ward and buffer rooms in the ward area, as well as operating rooms, ICUs, and CT rooms in the medical test and imaging area. The technology should also be utilized to conduct standardized design and develop the design into full-fledged products. That will contribute to the improvement of design quality and efficiency. Additionally, building, structure, electromechanical, decoration, equipment and other disciplines need to accept refined parametric modeling to facilitate the design of similar healthcare facilities.

II. Application of BIM Collaboration Platform

During the construction of the hospital, all buildings parties, operators and users adopt the visualization technology (Figure 3.6) and the cloud platform-based network collaboration platform, so as to display the results, collect opinions from all parties, and realize the real-time communication and coordination. Moreover, they can conduct coordinated management in progress, quality and cost, while giving more consideration to the need for intelligence operation (Figure 3.7) in the design-construction process.

III. Application of Digital Twin Hospital in Operation and Maintenance

Digital twin has two application dimensions: the geometric model that emphasizes physical characteristics; the management

Figure 3.6　Visualized Roaming of Leishenshan Hospital in BIM Model

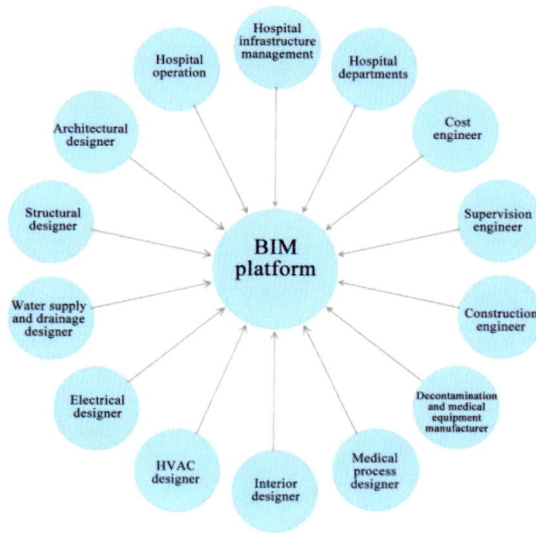

Figure 3.7　Information Transmission on BIM Platform during the Project

model that emphasizes digital application. Only by combining the two can it be brought into full play.

1. Medical Guide and Schedule Customization in Large Hospitals

According to its three-dimensional spatial attributes, the digital twin hospital provides medical guide and schedule customization on mobile terminals to help patients to "make registration, visit the hospital, go to the consulting room, accept treatment, and make payment" online, so as to improve their experience.

2. Whole-process Precise Operation and Maintenance of Building Facilities and Equipment

The digital, spatial and visual models of the building facilities provided by the digital twin hospital, are combined with the BA (building automation) system. It can monitor the energy use within the building in real time,

allowing timely and accurate troubleshooting and preventive maintenance of the equipment system.

3. Digital Management of Medical Equipment

Through the BIM model, the hospital can more accurately associate its space with CT machines, MR machines, inspection and testing machines and other asset-intensive medical equipment. It can store the parameters of maintenance equipment in the digital twin hospital. As a result, it can integrate and unify the virtual entities of equipment and text parameters and ensure the accuracy and timeliness of equipment asset management, which is conductive to the hospital asset management.

4. Big Data and Artificial Intelligence

Combined with artificial intelligence and the full operation simulation, the hospital collects the service data of medical space, such as the number of patients, treatment demand and time. It can adjust its space structure timely according to the change in demand, and can be pre-transformed in the digital twin hospital, and can give pre-judgment and pre-decision by virtue of big data and VR system.

3.1.4　Summary

Thanks to BIM technology, a vital support for design and construction, Leishenshan Hospital can complete its mission. Based on this, we will further carry out the research on building digital twin hospitals with BIM technology, so as to prepare for the coming intelligent era and emergencies.

3.2 Application of Prefabricated Building Design in Leishenshan Hospital

3.2.1 Overview

As an unexpected COVID-19 epidemic spreads rapidly in Wuhan, the Epidemic Prevention and Control Headquarters in Wuhan quickly decided to build Leishenshan Hospital. However, the Hospital, with a gross building area of 80 000 m² and 1 500 beds, must be completed in about 10 days. Receiving this order at a critical and difficult moment, Central-South Architectural Design Institute Co., Ltd. rushed to set up a design team, made technical planning, and quickly developed a construction proposal, in which an architectural design of prefabricated steel structure is applied.

Prefabricated building is a main green building promoted in current China. CSADI had been engaged in the design and research of prefabricated building for a long time and successively compiled many prefabricated building standards at national and provincial levels. It is also the main compilation organization of technical standards of China Association for Engineering Construction Standardization, such as *Specification for Design of Modular Prefabricated Monolithic Building* and *Technical Standard for Seismic Energy Isolation and Dissipation of Modular Prefabricated Monolithic Building* . The design of Leishenshan Hospital Project was rapidly completed in the context of such technical reserves.

Leishenshan Hospital was constructed based on the architectural design of prefabricated steel structure. Through technical planning, the integrated modular design was conducted on the structural system, building envelope system, MEP system and interior system of the prefabricated building according to the requirements of rapid industrial construction. Relying on BIM technology and the joint efforts of all participants, such a special project of global concern was completed successfully.

3.2.2 Technical Planning

Can the prefabricated building project be successfully implemented? In the initial design stage, it is critical to make technical planning together with the main participants of the project. Our design team attaches great importance to the technical planning, analyzes the overall characteristics, difficulties and key points of the project, and seeks for proper technical strategies through detailed analysis of influencing factors (Table 3.1).

Analysis of Project Characteristics   Table 3.1

S/N	Characteristics	Description
1	Extremely tight construction schedule	About 10 days（Normally, the design and construction duration should be more than 180 days）
2	Super-large scale	80 000 m², 1 500 beds, suitable for 2 300 medical staff
3	Complex building functions	A hospital for super infectious diseases, involving complex circulation of people flow and material flow
4	Site conditions	Good and suitable natural foundation
5	Project orientation	Driven by the "Chinese Speed" spirit, delivering an excellent project by utilizing all resources of China

S/N	Difficulties	Solutions	Measures
1	Complex building functions of emergency hospitals	Complex functional areas are subject to standardized and modular breakdown, and each module is designed separately	The standardized design is adopted for same or similar functional areas to facilitate construction
2	Complex multi-discipline cooperation	The integrated design is applied to multiple disciplines	Design data sharing through BIM platform
3	Extremely tight construction period to complete the design and decoration construction	The standard single-storey layout is adopted for some works. Design and construction race simultaneously against time	The standard modular breakdown of single-storey prefabricated building is fully considered to facilitate construction without installing the elevator
4	Large processing and procurement quantities of modular building	Different functional modules are subdivided and processed synchronously by category	The General Contractor searches in the whole country for suitable manufacturers and promotes production-synchronous procurement
5	Many construction staff on the site	The whole project should be based on single-storey construction and modular breakdown into several elements	The single-storey construction is carried out in sections synchronously to enlarge the operation area
6	Multi-procedure cross construction	The integrated modular design is used to reduce intermediate links such as the installation and decoration of main structure and equipment	The integrated modular prefabricated house is used, with each building function formed in the factory at one time. Overall installation and local adjustment are completed at the site

Through the above analysis, the difficulties in the project implementation are analyzed to search for solutions and measures (Table 3.2).

According to this technical planning, all disciplines and departments immediately began the stressful design and construction.

3.2.3　Technology Integration of Prefabricated Building

I. Architectural Design

The design is centered on the functional areas of the overall project according to the above technical planning. The building functions involve the isolation area, the living area for medical staff, and the comprehensive logistics area. The isolation area is integrated and the same modules are prefabricated according to the design concept of modular prefabricated building. The entire isolation area is arranged in a fishbone-like single-storey layout. All units can be divided into two basic units with a plane dimension of 3 000 mm × 6 000 mm and 2 000 mm × 6 000 mm respectively and a height of 2 900 mm. The module consists of the main structure, floor slab, wall board, ceiling, equipment pipeline, and built-in components.

The modules form a three-dimensional space with integrated functions. The test and imaging area holds the common prefabricated building with steel structure. The living area for medical staff includes dormitory area, office area, catering area and cleaning supplies warehouse. The dormitory area holds 10 dormitory buildings of 1- or 2-storey prefabricated steel structure of standardized design, with a building height of 7.5 m. Most of these buildings are located in the former stadium for the 7th CISM Military World Games and constructed from standardized light-steel prefabricated buildings featuring fast construction and mature technology.

II. Structural System

Modular prefabricated building with steel structure is a type of highly integrated prefabricated building involving much higher degree of industrialization than other types. Guangzhou University, Central-South Architectural Design Institute Co., Ltd., and other parties have compiled national or industrial standards for modular prefabricated buildings. Many Chinese steel structure manufacturers have developed and studied the processing technology of modular container house. CSADI has rapidly developed the modular prefabricated building design of Leishenshan Hospital based on the research and application of these technologies.

Since the building functions and spatial characteristics of the isolation area are different from those of the living area for medical staff, the modular prefabricated steel structure and the common prefabricated steel structure are adopted respectively in the design of Leishenshan Hospital (Figure 3.8, Figure 3.9) .

1. Modular Prefabricated Steel Structure

The care unit in the isolation ward and treatment area has the typical characteristics of standard modularization, so the building system of modular prefabricated steel structure is adopted.

This modular container unit is of steel structure frame and sandwich panels, and fea-

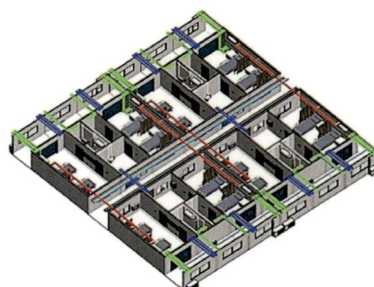

Figure 3.8　Ward Module Unit Assembly in Isolation Area 1

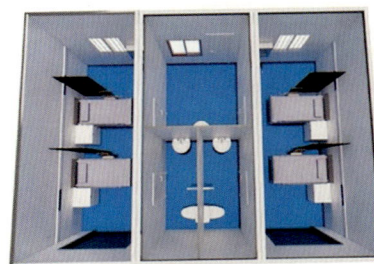

Figure 3.9　Ward Module Unit Assembly in Isolation Area 2

tures strong structural integrity, high bearing capacity, wind and seismic resistance performance, safety and serviceability. The modular container unit can be used alone, and also can be prefabricated and spliced freely according to the usage requirement, to form a large functional area through different combinations in horizontal and vertical directions. However, only the single-storey modules can be used in this project, and the total number of modules used is more than 3 190 (Figure 3.10) .

2. Common Prefabricated Steel Structure

(1) The test and imaging

area in the isolation area has a height of 4.3 m, and the plane column grid is not unified and the local span reaches 18 m, so it is necessary to use the common prefabricated steel structure building and the steel-framed structure (Figure 3.11, Figure 3.12, Figure 3.13) .

Figure 3.10 Field Installation of Care Unit Module Container House in Ward and Treatment Area

Figure 3.13 Field Installation in the Test and Imaging Area of Isolation Area

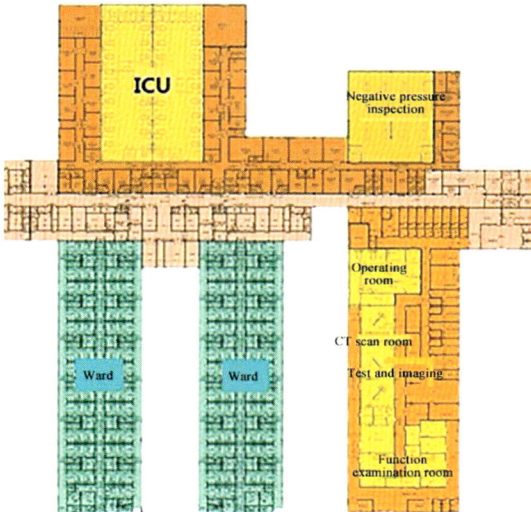

Figure 3.11 Local Layout of the Test and Imaging Area in the Isolation Area

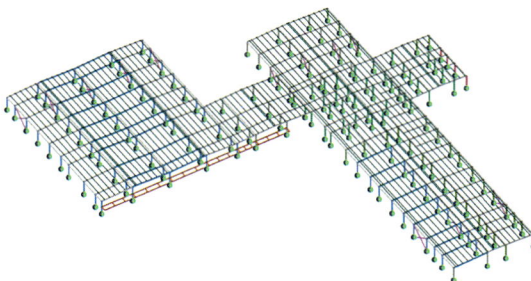

Figure 3.12 Structural Diagram of Test and Imaging Area (Midas Analysis Model)

（2）The living area for medical staff is constructed to hold two-story buildings with light-steel movable container house system according to the plane and space requirements of the building. The main structural unit is a light-steel structure frame. In order to ensure the horizontal resistance of the structure, cross stay cables are arranged between the frames to increase the structural lateral rigidity and safety.

（3）The light steel truss or H-shape steel beam is adopted for floor support system on which the structural slab is laid. The roof and wall enclosure are of color steel sandwich panels. With maturet echnologies and standardized modular production, the movable house is very convenient and fast to install and disassemble, and possesses obvious advantages as a temporary emergency building, which can meet the functional requirements of the building (Figure 3.14).

III. Building Envelope System

Two types of prefabricated buildings are

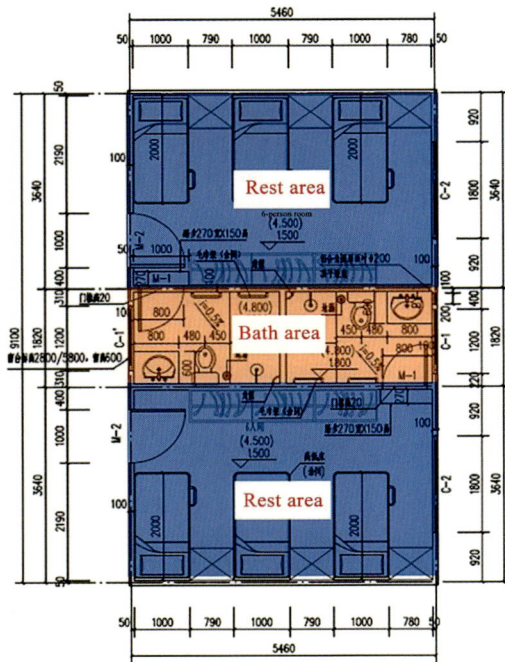

Figure 3.14　Typical Unit of the Living Area for Medical Staff

used for Leishenshan Hospital. For the modular prefabricated steel structure building, the building envelope system is prefabricated at the factory in accordance with the integrated design comprising the modular container house, structure, equipment, decoration, and others.

Leishenshan Hospital follows the general requirements of the building envelope system and stringent tightness requirements. A single modular container shall be prefabricated at the factory. The building envelope wall can meet the tightness requirements. However, it is required to cross the joints of the outer wall board of the container body when the container body is spliced into a larger bay. The gap between the container bodies is filled tightly with foam adhesive. The openings on the wall

board and the spliced joints are sealed with sealant and then covered with aluminum foil to ensure the air tightness.

The building envelope of common prefabricated steel structure is of rock wool composite board with profile dimensions of 1 820 mm × 950 mm × 100 mm ($L \times W \times T$). Rock wool composite board takes two-layer quality color-coated steel plate or other precision-pressed metal plate as the face plate, and the formed plate features excellent performance in fire protection, heat preservation, environmental protection and other aspects.

IV. MEP System

The design requirements on the MEP system of the prefabricated building are embodied in standardized design, equipment and pipeline module integration, and separation of the pipelines from the structure.

1. Standardized Design

The MEP system of this project is standardized and serialized through standardized design and the size of equipment pipelines, components and interfaces to ensure convenient procurement, strong versatility and interchangeability, and fast installation of components and parts.

(1) The standardized design and installation is adopted for water supply and drainage pipes. Indoor and outdoor water supply and drainage pipes (including drainage and ventilation) must be arranged separately at the ward and treatment area and other areas. The restrooms and isolation rooms at the isolation ward area are designed into a standard sanitary unit. The dormitory buildings for medical staff take the shower stall as the restroom, and the dormitory buildings for experts take the unit bathroom and others as the integrated restroom. Integrated design is adopted for wall (board) ceiling, sanitary equipment, pipeline and so on. *DN*50 rainwater pipes are arranged at four corners of the roof of the modular container house to drain the rainwater to the below-ground empty space of the container body.

(2) According to the HVAC design, more than 80% of ventilation and air conditioning ducts are of standardized PE pipes (Figure 3.15). The duct specifications are determined according to the division calculation of the system, and the diameters of the main ducts should be uniform. Main ducts are easy to install and only need a simple electric welding connection with branch ducts.

(3) The negative pressure isolation ward takes box-type centrifugal fan as the ventilation equipment (Figure 3.16). Before the equipment is delivered from the factory, the fan, motor and shock absorber have been integrated into a box, which can be quickly connected with the air duct on the site. The low-efficiency, med-efficiency and high-efficiency filters of the fresh air filtration system are also integrated into a high-efficiency filtration unit at the factory, without assembling of each part on site.

2. Modular Integrated Design

With the help of BIM technology, the modular integrated design is adopted for electromechanical equipment and pipelines and mainly features are as follows.

(1) Water supply tank is of prefabricated water tank as per the modular design, and all water is supplied by the break tank. Water supply and drainage equipment and pipelines are separated from the main structure to facilitate maintenance and replacement, during which the main structure shall not be affected.

(2) Modular design is adopted for the standard negative pressure isolation ward. In accordance with the requirements on the negative pressure isolation ward and the technical features of air distribution in the air supply and exhaust systems, the HVAC discipline has carried out CFD simulation (Figure 3.17), and identifies the accurate locations of indoor air

Figure 3.15   Pipeline Connection between the Roof-mounted Main Duct and the Ward

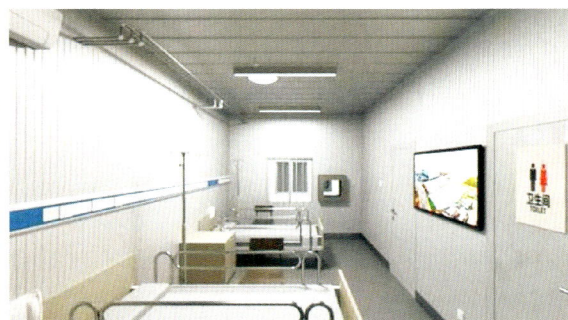

Figure 3.16   Rendering of BIM Design of Isolation Ward

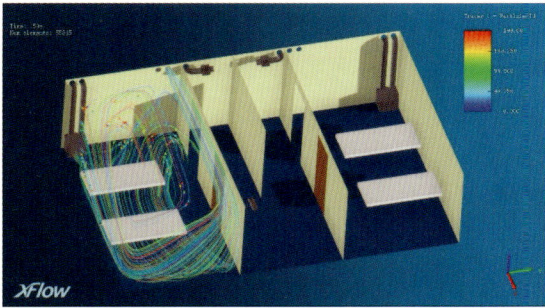

Figure 3.17　CFD Simulation of Flow Track in Ventilation Environment of Isolation Ward

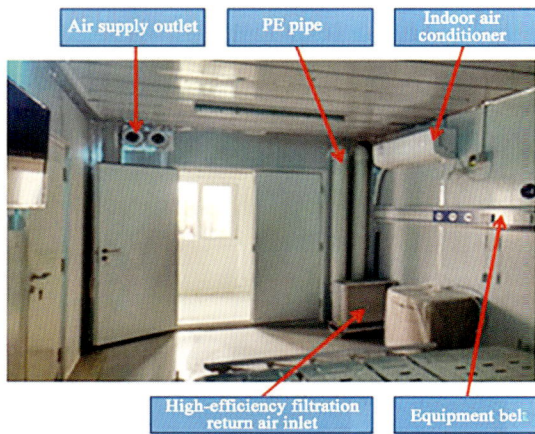

Air supply outlet　PE pipe　Indoor air conditioner

High-efficiency filtration return air inlet　Equipment belt

Figure 3.18　Equipment Installation of the Isolation Ward

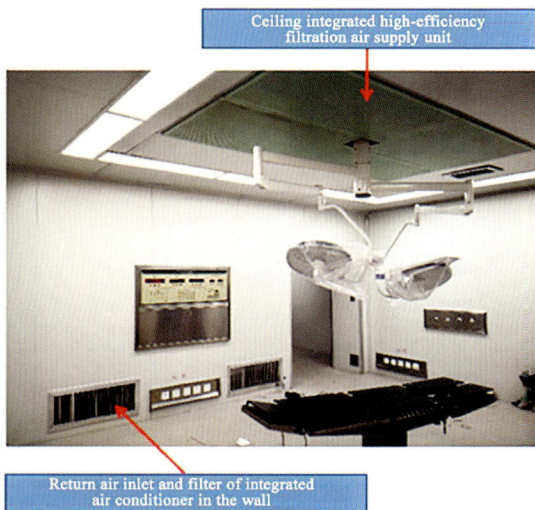

Ceiling integrated high-efficiency filtration air supply unit

Return air inlet and filter of integrated air conditioner in the wall

Figure 3.19　Integration of High-efficiency Air Supply Unit and Return Air Inlet in the Operating Room

distributor and high-efficiency air outlet based on simulation results. The equipment belt in the isolation ward is equipped with the integrated terminals for medical gases (e.g. oxygen, negative-pressure absorbed and compressed air), as well as the switches, power outlets and calling terminals (Figure 3.18).

(3) The negative pressure operating room is a unit module. In the operating room, the integrated E&M design is adopted for the air cleaning system, the air vent of the ventilation and air conditioning system, the medical gas settings and other facilities and equipment. The operating room is an independent module (Figure 3.19). The filtration and ventilation components are of prefabricated structure. The high-efficiency filtration air supply unit is integrated into the ceiling, and the return air inlet and filter of the air conditioner are integrated into the prefabricated wall. Cell-type air conditioner units are adopted, and the whole module of the operating room is prefabricated on site.

(4) Box-type substation and box-type silent diesel generator set are adopted. In order to facilitate construction and positioning and ensure the maintenance safety, they are processed and prefabricated in advance at the factory.

3. Pipeline Separation Design

Pipeline separation design is adopted for this project. Water supply and drainage, ventilation and air conditioning, electrical equipment and pipelines are separated from the main structure, and constructed with dry method. The main duct of ventilation and air conditioning of negative pressure isolation

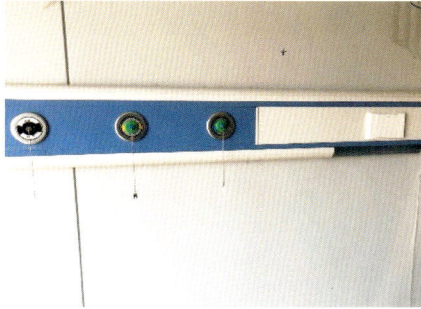

Figure 3.20 Terminal Integration of Equipment Belt in the Isolation Ward

ward is arranged on the roof of the container house, and the same of the test and imaging room is arranged in the ceiling of the public area. Fans and air conditioning equipment are installed on the roof or in the equipment room in the hygienic area. The medical gas pipeline in the ward is separated from the wall and installed in the equipment belt (Figure 3.20).

V. Interior System

The interior design of prefabricated building follows the principle of integration of building, decoration and parts. The industrialized interior parts are used to realize a complete set of integrated supply. The interchangeability and versatility of components and parts are improved through parameter optimization, tolerance fitting and interface technology. Decorative components and parts with variability and adaptability are convenient for construction and installation, service, maintenance and conversion. Modular prefabricated container house of Leishenshan Hospital can be put into service after integrated design and factory prefabrication of building, structure, decoration, components and parts, field installation, and proper local adjustment. This plays an important role in the timely completion of the hospital project. The

envelopment plates of the common prefabricated steel structure building in the test and imaging and living areas are also industrial plates integrated with thermal insulation and decoration functions, which are convenient for installation and connection. In each area of the hospital, the restroom is installed with industrial components and parts, and the shower cabin is applied. Most of the electromechanical equipment achieves integrated supply of complete sets. These parts and components are easily installed on the site, which makes it possible for the completion of the 80 000 m^2 prefabricated building within an extremely tight construction schedule.

VI. Information Management, Prefabrication and Installation

The digital simulation is applied in the design and construction of prefabricated buildings, contributing to improving the design and prefabrication efficiency. BIM technology is applied to the integration and classification of buildings, structural components and electromechanical equipment, to directly guide the factory prefabrication. In addition, the BIM technology is used to simulate the site construction by virtue of its stimulability and identify the best assembly scheme, thus greatly improving the assembly efficiency and speeding up the construction. The time for design and construction of Leishenshan Hospital is too short to complete the BIM-based full-model forward design, but the designers cooperate with the steel structure manufacturer to directly exchange the data of design model and factory prefabrication when the main structure of prefabricated steel structure in the test and imaging area is designed.

The construction drawings are taken as detailed manufacturing drawings and directly imported to workshop production, thus saving a lot of time.

Due to the tight schedule and urgent heavy task, the design, construction, prefabrication and installation parties effectively cooperate and coordinate with each other. The design follows the principle of standardization, modularization and integration, and uses the mature industrialized product system as much as possible. During the construction, many alternative methods are adopted as actually required, such as replacing the original empty-space structure with Bailey beam. In addition, problems of column base installation are solved. In this process, the advanced BIM technology is utilized to solve the contradiction of the cross-construction of this project.

3.2.4　Summary

（1）The integration of building, structure, equipment and interior decoration comes true in a fast and efficient manner through innovative technical planning, the modular prefabricated steel structure building model and the advanced technology integration. The successful application of innovative technologies in Leishenshan Hospital fully demonstrates the high application value and huge development potentials of modular prefabricated building system.

（2）Flexible alternatives for structure and foundation have earned valuable time for the successful implementation of the whole project.

（3）The standardization, modularization and pipeline separation design of MEP system

of prefabricated building has been successfully applied in Leishenshan Hospital and fruitful results were achieved.

（4）The integration of design, construction, prefabrication and installation of the prefabricated building is just the development trend of transformation and upgrade of construction industry in China, and the EPC mode combining design and construction can ensure the efficient implementation of the project. The integrated design and construction of Leishenshan Hospital fully shows that this model is full of vitality.

3.3　Application of Numerical Simulation

Customized design is needed to meet the temporary epidemic prevention requirements of the ventilation system of Leishenshan Hospital. The special air system for negative pressure ward was adopted in the design of the hospital, which is capable of effectively prevent virus from spreading into the hospital, avoiding cross infection due to virus spreading. In the negative pressure isolation ward, a reasonable ventilation system can effectively not only improve the air quality in the ward, but also contribute to patient rehabilitation, and protect medical staff to the greatest extent. In addition, Leishenshan Hospital is also provided with a special system to exhaust polluted air out of the ward aloft, which is in favor of reducing the risk of pollution to other areas around the hospital.

With traditional design methods, it's

neither possible to quantitatively compare different design options regarding the concentration of polluted air in the wards of Leishenshan Hospital, nor to evaluate the impact of emission of such polluted air on external environment around the hospital. No matter concerning the construction of ventilation system in wards, the initial site selection of hospital, or the prevention and control of secondary pollution of exhaust emission during hospital operation, an analysis method which can effectively simulate the diffusion mechanism of polluted air is urgently needed to provide more quantitative and specific reference and basis. Moreover, considering the needs of Leishenshan Hospital, a temporary hospital with a short design and construction period, the chosen digital analysis method requires not only sufficient calculation accuracy, but also higher calculation efficiency. The diffusion trajectory and special concentration distribution of polluted gases can be quantitatively analyzed with Computational Fluid Dynamics (CFD) technology, providing a reference to the design and site selection of the temporary hospital. Among numerous CFD analysis software XFlow, the software based on Lattice Boltzmann Method (LBM), has the advantages of both macro-fluid continuous model and micro-analysis dynamics method, and features of high computational efficiency and satisfactory parallelism.

This section mainly introduces the application of XFlow-based CFD technology in the design of the ventilation system for Leishenshan Hospital. With the cutting-edge CFD software XFlow, the airflow distribution and pollutant diffusion in negative pressure wards with several layout options of the ventilation system was compared, to find the best option with least pollutant diffusion and to protect medical staff and prevent cross infection to the greatest extent. In addition, an urban-level large-scale flow field model is established to study the impact of exhaust emission from hospital wards on the surrounding environment, which provides a reference to the overall planning and design of the hospital and for the health and safety of medical staff, patients, and the public.

To study the impact of air supply and outlet layout of the ventilation system in wards of Leishenshan Hospital on the diffusion of polluted air, four different air conditioning systems are taken into consideration (Figure 3.21 to Figure 3.25). The Rhino software with efficient parametric modeling function is adopted to establish the multi-variant models of the negative pressure isolation ward, facilitating the option comparison of the ventilation system. Comparison of the ventilation system options given in Table 3.3 indicates that, the negative pressure isolation ward specified in option A has lower concentration of polluted air and higher ventilation efficiency.

In order to study the impact of polluted air exhausted from Leishenshan hospital on surrounding crowds, the conditions of northeast wind (blowing towards the residential area) and southeast wind (blowing towards the living quarter of the hospital) are selected for diffusion analysis of polluted air (Figure 3.26). Un-

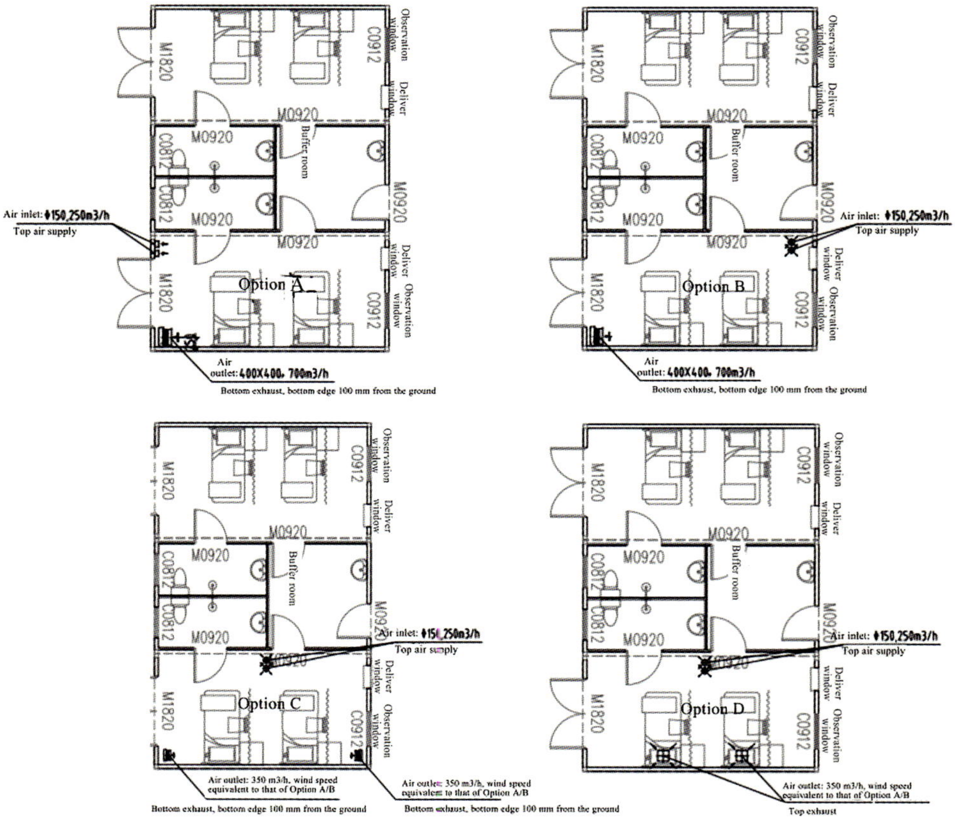

Figure 3.21　Four Different Air Supply and Exhaust Options

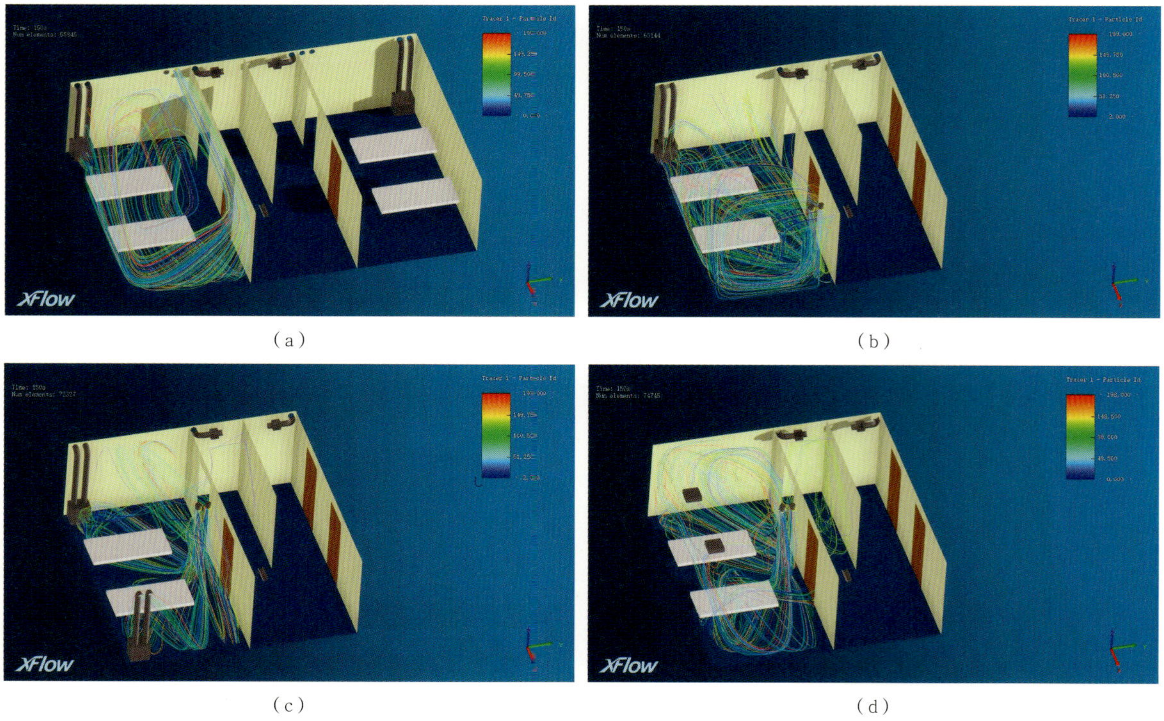

Figure 3.22　Comparison of Airflow Trajectories in Four Options
（a）Option A；（b）Option B；（c）Option C；（d）Option D

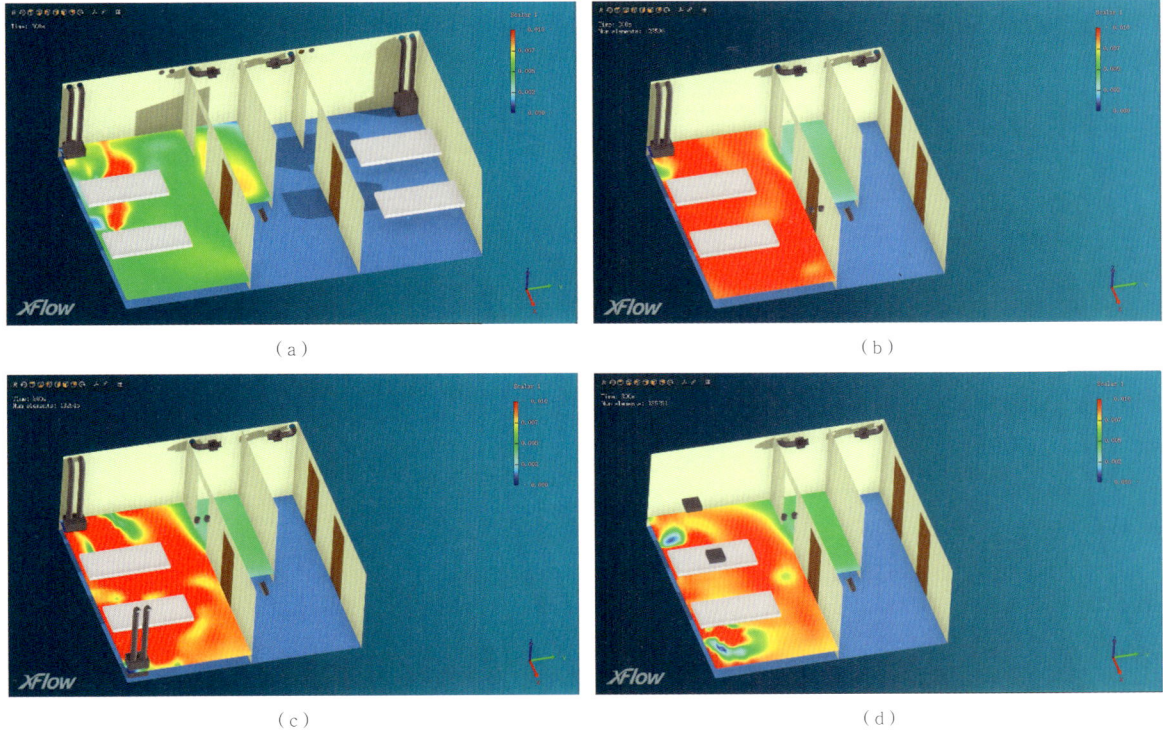

Figure 3.23 Concentration Comparison of Four Options — Exhaust Point Height
（a）Option A； （b）Option B； （c）Option C； （d）Option D

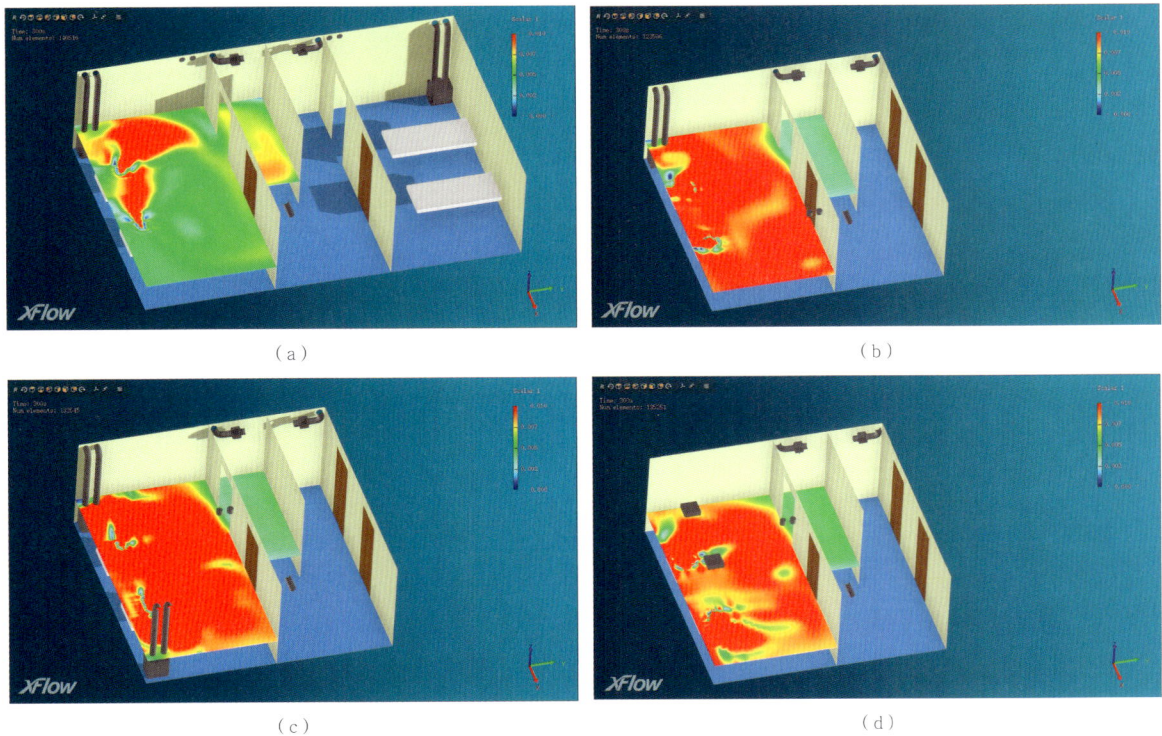

Figure 3.24 Concentration Comparison of Four Options — Patient Face Height
（a）Option A； （b）Option B； （c）Option C； （d）Option D

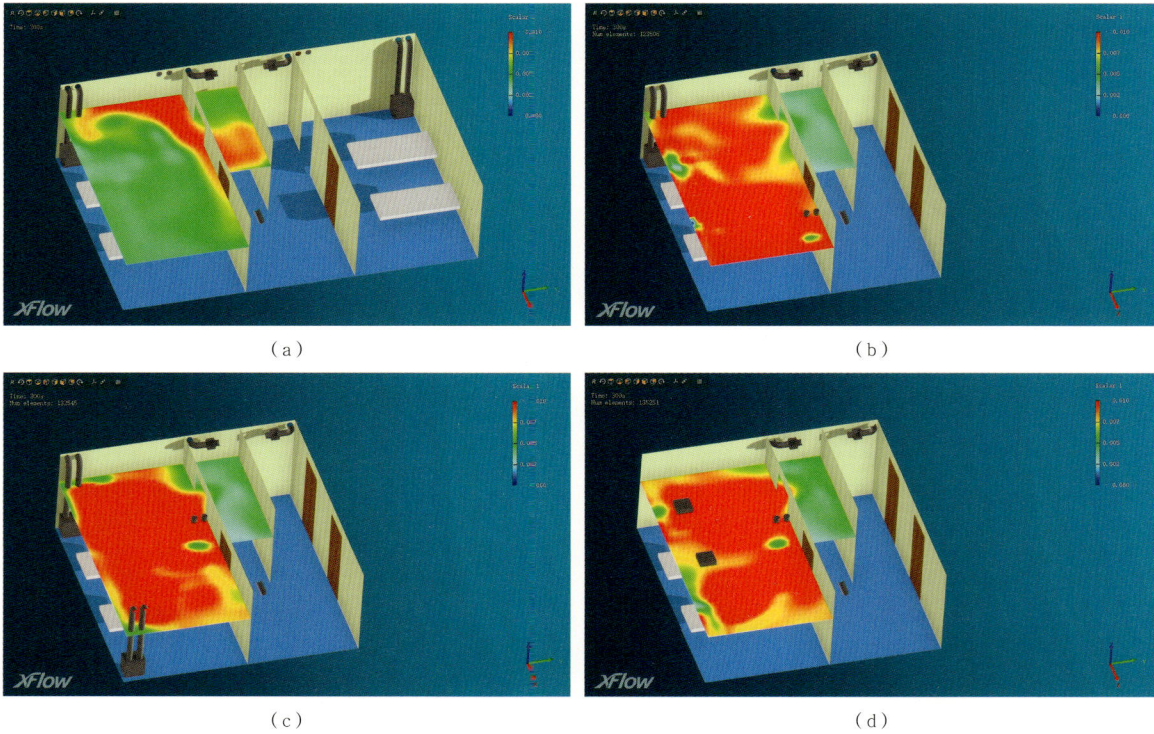

Figure 3.25 Concentration Comparison of Four Options — Doctor Standing Height
（a）Option A; （b）Option B; （c）Option C; （d）Option D

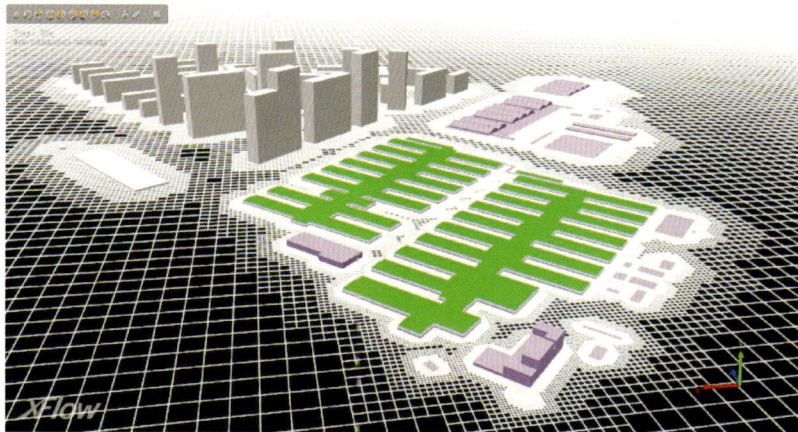

Figure 3.26 Urban-level Large-scale Flow Field Analysis Model of Leishenshan Hospital

der the northeast wind condition, the polluted air in the residential area is diluted to over 43 478 times, as shown in Figure 3.27 （a）. Under the east wind condition, the polluted air in the living quarter of the hospital are diluted to over 169 000 times, as shown in Figure 3.27 （b）.

According to related research, SARS virus is no longer infectious when it is diluted to more than 10 000 times. Therefore, it is believed that Leishenshan Hospital has less influence on other areas and public safety will not be affected.

Design and Construction Technology of Digital Emergency Hospital

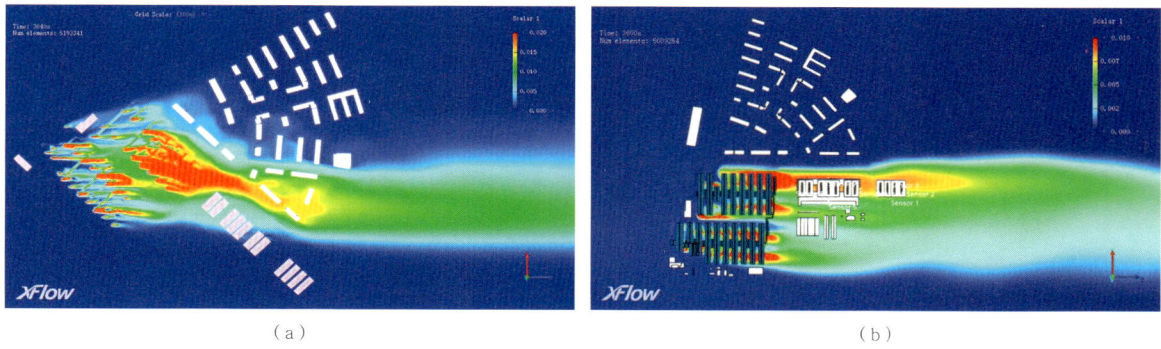

（a）   （b）

Figure 3.27　Working Condition under Northeast and East Wind

（a）Working Condition under Northeast Wind；（b）Working Condition under East Wind

Concentration Comparison of Four Options

Table 3.3

Ventilation Option	Exhaust Point 1	Patient 1 Point 2	Patient 2 Point 3	Doctor 1 Point 4	Doctor 2 Point 5	Average in Ward
Option A	0.006 9	0.005 2	0.007 7	0.004 4	0.004 6	0.006
Option B	0.008 2	0.069 1	0.166 1	0.009 7	0.014 9	0.009
Option C	0.008 7	0.006 8	0.012 8	0.009 7	0.013 8	0.011
Option D	0.008 6	0.002 4	0.162 2	0.011 2	0.023 1	0.009

附录

医疗建筑精选

华中科技大学同济医学院附属同济医院光谷同济医院项目

1. 项目概况

（1）建设地点：湖北省武汉市。

（2）建设规模：用地面积 166448m²；建筑面积 174719.6m²。

（3）建设情况：一期已完成，二期建设中。

（4）业主单位：华中科技大学同济医学院附属同济医院。

（5）获奖情况：全国优秀工程勘察设计建筑工程三等奖。

2. 项目简介

光谷同济医院项目以优化患者就医体验为设计主旨，以患者为中心，为患者创造"家"的环境，以及便利、舒适的空间。住院部的入口大厅引入酒店大堂式的服务理念，使患者一进医院就感受到亲切轻松的气氛，既方便患者就医，也减轻患者的心理压力。

在设计中体现医疗建筑特征的同时，强调与周边的城市肌理相协调，强调内部功能美与外在形式美的有机结合，将表现医疗建筑特点作为设计的基本构想。采取半集中式的布局模式，通过南北向的医疗街，将门急诊、医技、住院空间较为紧凑地组合在一起，形成有机的城市医疗空间。医疗街周围舒展的造型结合多角度偏转的朝向，为患者创造更丰富、多维的景观资源，营造自然轻松的氛围。

同济医院光谷院区

华中科技大学同济医学院附属协和医院外科病房大楼

1. 项目概况

（1）建设地点：湖北省武汉市江汉区解放大道 1277 号。

（2）建设规模：本项目是以 640 张病床、45 间手术室为主的综合性外科病房大楼，总建筑面积 74117m²，地下室建筑面积 8888.6m²，其中裙楼 6 层，主楼 32 层，地下 2 层，建筑高度 127.8m，最高点 144.7m。

（3）建设情况：已建成。

（4）业主单位：华中科技大学同济医学院附属协和医院。

（5）获奖情况：2007 年湖北省勘察设计行业优秀建筑工程设计一等奖；2008 年度全国优秀工程勘察设计行业奖建筑工程三等奖；2007 年度中国建筑工程鲁班奖（承建商）；2007 年优秀建筑结构设计二等奖；中国建筑学会建筑设备优秀设计二等奖（给排水专业）；中国建筑学会暖通空调工程优秀设计二等奖。

2. 项目简介

项目设计充分遵循"以患者为本"的指导原则，将人流量大的科室布置在低层。根据弧形的建筑形体调整各功能空间尺寸，形成既便于使用又富有空间趣味的医疗诊疗、护理单元。病房皆沿弧形朝南布置，并依托造型及医疗需求调整病房尺寸，让每位患者都能享受和煦阳光和开阔视野，以利身体康复。医疗服务、后勤供应、垂直交通系统皆置于平面北侧，形成相对独立的区域，避免干扰病房。利用中部空间较宽处布置护士站，两头狭窄空间布置辅助功能，给患者提供便捷的服务。从而形成空间利用非常合理，又别具一格的医疗综合住院建筑。

外科病房大楼

华中科技大学同济医学院附属协和医院门诊医技大楼项目

1. 项目概况

（1）建设地点：湖北省武汉市江汉区解放大道以北，新华路以西，解放大道 1277 号。

（2）建设规模：地下 3 层，地上 23 层，建筑面积 87780m^2。

（3）建设情况：已建成。

（4）业主单位：华中科技大学同济医学院附属协和医院。

（5）获奖情况：2011-2012 年度湖北省建筑工程安全文明施工现场（楚天杯）；2012-2013 年度湖北省建筑结构优质工程；2013 年度建筑工程黄鹤奖金奖；2013-2014 年度（第二批）湖北省建设优质工程（楚天杯）；2014 年湖北省建筑节能示范工程；2014 年度湖北省优秀工程设计二等奖；2014-2015 年度第一批中国建设工程鲁班奖（国家优质工程）。

2. 项目简介

华中科技大学同济医学院附属协和医院门诊医技大楼项目位于武汉市江汉区解放大道 1277 号，设计对象为集门诊、手术、医技、教研、病房、地下车库等功能空间于一体的综合大楼。地下二层作为变电所和停车库，和医院已有外科楼地下室相连，与武汉地铁 2 号线中山公园站无缝连接，较大地提高了经济效益，同时使得医院整体规划更加完整，各功能区如门诊楼、外科楼、内科楼、保健楼的联系更加紧密、便捷，同时使得城市的建筑空间和景观更加丰富和完整。

服从城市规划整体布局要求，严格控制建筑密度及绿化率。建筑物严格按规划要求后退道路红线，保证了城市空间的连续性和完整性。

以患者为本，合理布局。根据项目组成空间的不同功能要求，合理划分及布置功能区域，最大限度地发挥环境资源优势及经济效益。

门诊医技大楼

华中科技大学同济医学院附属同济医院内科综合楼项目

1. 项目概况

（1）建设时间：2015 年 5 月 ~2019 年 2 月。

（2）建设规模：用地面积 10780m² ；建筑面积 103565.48m²。

（3）建设情况：已建成。

（4）床位数：1198 张。

2. 项目简介

华中科技大学同济医学院附属同济医院内科综合楼地处院区西北角，设计注重院区整体规划，通过项目的实施与已有建筑在院区中部围合出大面积开敞景观庭院，改善院区整体环境。遵循以人为本的设计理念，在交通空间中引入景观庭院和空中花园，底层布置咖啡吧、茶吧等休闲场所，使得医院不仅在生理上为患者治疗，也在心理上给予患者温暖的慰藉。

项目用地不规则，设计有效利用现有地形布置出适合医疗功能要求，并具有良好通风与景观的平面布局，得到业主充分认可。设计采用了多样的遮阳构件，在保证采光通风效果的同时，遮挡武汉夏季强烈的直射日光，同时也使建筑更具有细节感、更加精致，体现现代医疗建筑的特点。

内科综合楼

湖北省中医院国家中医临床研究基地湖北项目

1.项目概况

（1）建设地点：湖北省武汉市。

（2）建设规模：占地面积2万平方米；建筑面积6万平方米。

（3）建设情况：已建成。

（4）业主单位：湖北省中医院。

2.项目简介

本项目地处湖北省中医院光谷院区，位于武汉市洪山区珞瑜路856号。国家中医临床研究基地湖北项目是一座国家级中医临床研究基地大楼，是湖北省中医院进行包括肝病等3个重点学科、6个重点专科临床研究和科研建设的重要基础平台。为进一步提升湖北省中医院的临床研究能力，不断推进中医药的科研水平，本项目设计时除遵照相关规划及建筑规范外，还结合项目独特的条件，使其契合中医临床研究发展的时代要求，力图使湖北省中医院国家中医临床研究基地成为武汉市重要的城市景观建筑及地域性地标。

沿街人视图

湖北省肿瘤医院门诊综合楼项目

1. 项目概况

（1）建设地点：湖北省武汉市洪山区。

（2）建设规模：占地面积 1.8 万平方米；建筑面积 3.3 万平方米。

（3）建设情况：已建成。

2. 项目简介

医疗建筑的形象要突出自身的个性特征，其外观也是其功能的空间体现，为创造绿色医疗环境，设计强调内部功能美与外在形式美的有机结合。设计注重空间的阳光感、流动感和体量感，在细部处理上充分体现了材质的轻重、虚实的对比。

为呼应医院滨临南湖、自然生态的特点，设计中运用曲线元素使得建筑更加柔和舒展。建筑内部空间开敞明亮，增加建筑的亲和力与开放性，注重标志性而不失典雅，细部丰富而有构造脉络可寻。以浅色系的外装饰面材料结合浅灰色透明玻璃窗，力求反映医疗建筑整洁和素雅的行业特色。

平面南北向呈 E 字形布局，组成立体式院落空间，保证了建筑内部房间享受自然通风采光的均好性。通过层数错落形成屋顶绿化，极大改善了就诊及办公环境。

门诊综合楼鸟瞰图

湖北省妇幼保健院洪山院区项目

1. 项目概况

（1）建设地点：湖北省武汉市。

（2）建设规模：用地面积 32625.48m²；建筑面积 155540m²。

（3）建设情况：建设中。

（4）床位数：500 张。

（5）建设单位：湖北省妇幼保健院。

2. 项目简介

湖北省妇幼保健院洪山院区依照"智慧医谷，孕育新生"的理念设计，建筑整体造型犹如一位母亲环抱爱子，场景温暖。医院设有门诊医技楼、住院综合楼，重点打造妇保、小儿外科、儿童神经康复三大医疗中心。门诊医技楼和住院综合楼内分别设置医技中心，考虑孕产妇、儿童、保健人群的就医、待产动线，最大限度地降低普通患者和健康保健人群动线交叉的可能性。

基地北侧由南向北布置门诊医技楼、住院综合楼。总体形成具备 500 张病床的功能完备的妇幼保健医院，各功能区域之间联系紧密又可高效独立运作。

高度上采用南低北高的建筑布置方式，利用高层和裙房的组合形成起伏有致的城市界面。南侧退让出面对城市的景观广场，为就诊人群提供良好的就诊、康复环境。

人视效果图

武汉光谷国际医院规划设计方案

1. 项目概况

（1）建设地点：武汉市。

（2）设计时间：2018 年 8 月。

（3）建设情况：方案阶段。

（4）建设规模：用地面积 9.7 万平方米；建筑面积 30.3 万平方米。

（5）设计单位：中南建筑设计院股份有限公司；夏邦杰建筑设计咨询（上海）有限公司。

（6）获奖情况：2020 年第五届中国十佳医院建筑设计方案。

2. 项目简介

本项目选址于武汉东湖新技术开发区新行政办公中心所在地。地理位置优越，交通便利，周边环境地形起伏、水网环绕，自然环境优美。项目定位为具有国际影响力，拥有大专科、小综合、多中心酒店公寓等多种功能，集多种元素于一体的示范性新型医院，光谷区独具特色的城市新地标。

建筑在基地的 4 个地块均衡设置，采取合院式布局，整体空间错落有致，既统一又富于变化性，开放空间结合北面的城市绿带与地铁上盖空间，形成脉络清晰的空间廊道，与四组建筑遥相呼应，创造出武汉光谷国际医院独具特色的城市空间，形成立体多变的城市天际线。

总体规划鸟瞰图

广东医学院附属医院海东院区项目

1. 项目概况

（1）建设地点：广东省湛江市海东新区。

（2）建设规模：建筑面积 47 万平方米。

（3）建设情况：建设中。

（4）床位数：3000 张。

2. 项目简介

广东医学院附属医院海东院区拟建设成为一座规模达 3000 张病床的，以急危重症医疗中心和专科医疗中心为功能定位的"大专科、强综合"综合性医院。

设计以人为本，以医患为中心，遵循"人性化、园林化、智能化、信息化、环保节能、可持续性发展"的原则。规划构架采用国际先进的医疗中心组团式布局，形成两轴、四组、六园，具有城市化与开放性。穿插大小多处庭院，采用生态布局和多项节能技术措施，创建绿色温馨的"医疗之家"。

这是一座由诊疗中心组团加医技平台组成的医疗综合体，建筑如同一艘医学航母，各诊疗中心如同旗舰群。建筑如生命航船，扬起风帆，承载着整个医疗城昂首向前，又如群鸟张开双翼，拥抱"绿"的海洋，体现海东新区"一湾两岸"和滨海城市医疗建筑的独特风采，昭示广东医学院附属医院海东院区必将迈向光辉灿烂的美好未来。

总体规划鸟瞰图

海南省人民医院秀英门诊楼及内科楼项目

1. 项目概况

（1）建设地点：海南省海口市。

（2）建设规模：规划总用地面积 213700m²。本工程为院区一期工程，用地面积 24380m²。总建筑面积为 90816.13m²，地下室面积为 26680.2m²。

（3）建设情况：已建成。

（4）业主单位：海南省人民医院。

2. 项目简介

本项目由 5 层的门诊楼、17 层的内科楼及其之间的连廊组成。

本项目门诊楼横向医疗街的设计使门诊楼和内科楼与已建成院区巧妙结合，既解决了前广场空间不足的问题，又使各功能布置清晰合理，交通组织顺畅高效，大幅提升院区空间品质。另外，根据海南湿润的气候特点，将医疗街的顶棚设置成膜结构遮阳屋面，配合庭院、天井和立面的多处电动通风百叶，使医疗街具有良好的通风和采光。

沿街鸟瞰图

河南省疾病预防控制中心项目

1. 项目概况

（1）建设地点：河南省郑州市郑东新区。

（2）建设规模：总用地面积 42500m^2；建筑面积 34190m^2。

（3）建设情况：已建成。

（4）业主单位：河南省疾病预防控制中心。

2. 项目简介

河南省疾病预防控制中心位于河南省郑州市，由 4 栋建筑组合而成。北边 2 栋为后勤楼及管理办公楼，高 8 层，前部由连廊相连，后部设架空天桥，并设有 1 层地下室。南边 2 栋均为实验楼，高 6 层，前部由连廊相连，后部独立设 P3 实验室。两组建筑之间设主门厅及会议中心。

沿街立面图

援苏里南共和国瓦尼卡医院项目

1. 项目概况

（1）建设地点：苏里南共和国瓦尼卡区。

（2）建设规模：用地面积 59000m² ；总建筑面积 13500m²。

（3）建设情况：已建成。

（4）床位数：180 张。

（5）获奖情况：中南院优秀方案（青年组）二等奖。

2. 项目简介

援苏里南共和国瓦尼卡医院位于苏里南共和国首都帕拉马里博以南的瓦尼卡区，距首都约 22km。由于目前苏里南共和国的医院大多集中在首都，瓦尼卡区以及周边的三个区都没有医院，周边地区居民看病非常不方便。瓦尼卡医院不仅为瓦尼卡区服务，而且将为瓦尼卡周边三个区的居民提供医疗服务。

瓦尼卡医院用地面积约为 100000m²，南北长度约为 526m，东西长度约为 190m。本次项目规划设计用地面积 59000m²，床位数为 180 张，其中包含 80 张病床的护理病房、100 张病床的低护理病房。总建筑面积约为 13500m²，日门急诊人数大约为 150 人。医生、护士及其他人员约为 200 人。

总体规划鸟瞰效果图

武汉市普仁医院项目

1. 项目概况

（1）建设地点：湖北省武汉市青山区。

（2）建设规模：占地面积1.8万平方米；建筑面积7.2万平方米。

（3）建设情况：已建成。

2. 项目简介

立足医院整体发展，强调整体性、协调性与可持续发展，为创造绿色医疗环境，整合了医院的空间，结合医院得天独厚的地理位置，将院区定位为城市医疗公园。作为和平公园的延续，公园、医院互为彼此的后花园，在医廊及医疗单元中引入景观庭院和空中花园，并布置不同层次的休憩、游园场所，给医疗工作者和患者带来勃勃生机与阳光心态。

新建医技综合大楼在造型设计上与已建门诊综合大楼形成充分的呼应，采用求同存异的细节变化，增强院区形象的统一感受。

在有限的用地范围内，优化布局、合理规划，保证医院持续高效运营。对道路、停车区域进行规划整合，配以绿化小品，形成医院内部的功能型庭院。

鸟瞰效果图

武汉市东西湖区人民医院异地新建项目

1.项目概况

（1）建设地点：武汉市东西湖区。

（2）建设规模：总用地面积 102297.3m²，一期建设总建筑面积 130000m²，地上总建筑面积约 108000m²，地下总建筑面积约 22000m²。

（3）建设情况：已建成。

（4）床位数：1000 张。

（5）业主单位：东西湖区人民医院。

（6）获奖情况：2018 年度中国医院建设匠心奖医院建设年度优秀项目。

2.项目简介

遵循"以人为本"的设计原则，以患者为中心，以改善就医环境为设计思想。本项目采用中轴鱼骨式的布局方式，根据项目用地特质及周边用地性质，形成贯穿南北的 2 条曲线形体控制轴，以此控制医院的总体布局，并依此设置景观，与西面的城市景观绿化廊道、东面的城市商业街区有了呼应与退让。不仅形成了清晰的布局模式，也积极运用了城市肌理，建立了良好的城市形象。

为医护人员创造单独使用的护理路线与环境良好的区域，使医患适当分流，减少交叉干扰。设置单独运营的污物系统，与此同时，供应系统结合医生交通流线设置，使这座医疗综合体的物流运营快捷高效。充分利用地形特点，结合周边环境优势，建设绿色节能医院，降低医疗综合体运营成本。

总体规划鸟瞰效果图

三亚市榆红医院项目

1. 项目概况

（1）建设地点：海南省三亚市。

（2）建设规模：总建筑面积 5.6 万平方米。

（3）建设情况：建设中。

（4）业主单位：三亚市榆红医院。

2. 项目简介

作为一家三级精神专科医院，建筑造型应能体现精神疾病医院的特点，给患者以平和温馨的心理感受，从此出发，设计结合功能需要，创造出独特的建筑形象和空间感受。

建筑整体形象如同展开的双臂和呵护心灵的双手，希望能给就诊者传递友好亲切的形象感受，舒缓就诊者和家属的心理压力，营造轻松温馨的就诊氛围。建筑造型语言以柔和平缓的白色和深蓝色相间的水平曲线为主，犹如舒缓的轻音乐，又如平静的海波，使整个院区沉浸在轻柔温馨的气氛之中。

建筑功能空间围绕一个半室外的中心庭院展开，把新鲜的空气、微风和漫反射的自然光线引入各个角落，立体化的绿植渗透于各个空间，营造出关爱和治愈心灵的港湾。

本项目是继海南省人民医院门急诊楼与内科楼、海南省儿童医院、三亚农垦医院之后，中南建筑设计院海南华筑国际工程设计咨询管理有限公司中标的又一个大型专科医疗项目。本项目的中标，增强了中南建筑设计院海南华筑对大型医疗建筑专业设计的信心，对公司开拓大型医疗项目、深度研究各种类型的医疗建筑设计市场具有重要意义。

鸟瞰效果图

武汉市中心医院北院区项目

1.项目概况

（1）建设地点：湖北省武汉市。

（2）建设规模：用地面积 2.4 万平方米；建筑面积 11.7 万平方米。

（3）建设情况：已建成。

（4）床位数：1200 张。

（5）业主单位：武汉市中心医院。

2.项目简介

规划设计时，武汉市中心医院改扩建工程用地位于武汉市江汉区姑嫂树路 14 号，姑嫂树路以东，规划支路以北，南侧紧靠拆迁区，北侧紧邻武钢集团汉口冷轧有限责任公司。

本项目用地基本为矩形，东西向长约 105m，南北向长约 220m。强调了资源整合与集中管理，使土地得到了最大化利用。尽量不影响院区内已有的医疗营运功能，将加速器楼完整保留。新建大楼靠近道路设置，并将门诊、急诊、医技、住院空间整合为一个整体，通过医疗街贯穿所有功能空间。建筑周边留出环形道路，很好地满足了交通和消防的需要。建筑在两个街角处后退形成广场，营造了环境优雅的入口空间，强调了建筑的入口。街角广场的设置也照顾了建筑与城市的关系。以医疗街为核心和纽带，各功能空间沿医疗街交通轴线展开，形成水平与垂直生长的模块，突出医院结构的清晰和可识别性。室内空间绿化向外部环境延续，内外相和，体现人与自然共存的生态概念，为患者营造充满希望的空间氛围。

沿街效果图

襄阳市中医医院东津院区项目

1.项目概况

（1）建设地点：湖北省襄阳市。

（2）建设规模：用地面积约 21.4 万平方米；总建筑面积 26.8 万平方米。

（3）建设情况：建设中。

（4）床位数：1000 张。

（5）业主单位：襄阳市中医医院。

（6）获奖情况：中南院优秀方案二等奖。

2.项目简介

襄阳市中医医院东津院区项目位于襄阳市东津新区团结坝水库以西、内环线东南角，西接东津大桥，北达东津高铁站，用地分为南北两个地块，规划总用地面积 214151m²，欲将襄阳市中医医院（中医药研究所）东津院区规划为扎根鄂西北本土、辐射中部地区、面向未来的中医医疗综合体。

总体规划鸟瞰效果图

荆州市中心医院荆北新院项目

1. 项目概况

（1）建设地点：湖北省荆州市。

（2）建设规模：占地面积 13.3 万平方米；建筑面积 25.8 万平方米。

（3）建设情况：建设中。

（4）床位数：2500 张。

（5）业主单位：荆州市中心医院。

2. 项目简介

规划时，荆州市中心医院荆北新院项目地处荆州市荆北新区核心区域，南邻楚源大道，北靠怀沙路，西为翼德路，东为郢南路，拟完成 2500 张病床的大型综合医院的建设规模，分两期实施。

设计采用以诊疗中心加医技平台的医疗综合体设计理念，以影像中心、检验中心、中心手术部等主要医技科室组成医院共享平台，以此为轴心，其他诊疗单元围绕此技术平台展开，这种功能布局保证了各医疗单位运行的高效性和独立性。采用医疗中心式布局模式，以人体系统器官为诊疗目标，各中心通过医疗街与医技平台紧密相连，其中大型优势专科均独立设置出入口。

建筑群体采用简洁的体量组合、现代的设计方法，整体形象统一大气，与环境完美融合。建筑立面追求细节的刻画，展现了现代医疗技术美学高精化、服务人性化的前瞻愿景。

总体规划鸟瞰效果图

随州市中心医院文帝院区项目

1. 项目概况

（1）建设地点：随州市经济开发区。

（2）建设规模：总用地面积约为 191 亩（约 127333m²），总建筑面积 22.9 万平方米，其中地上建筑面积 18.9 万平方米，地下建筑面积 4.0 万平方米。

（3）建设情况：已建成。

（4）床位数：2000 张。

2. 项目简介

随州市中心医院文帝院区项目是整体迁建项目，包括门诊、急诊、医技、住院、传染病、行政、教学、学生宿舍、保障用房、食堂、地下车库、室外工程等。建成后病房拟安置病床 2000 张，是一座大型综合教学医院。

设计时基于"功能模块化，流程体系化"这一全新的医院设计理念，坚持以患者为中心的同时，改善医护人员的工作和卫生安全条件，精心打造一座全新的现代化医院，为社会服务，为人民服务，在炎帝神农故里、编钟古乐之乡的医院建设业绩上，画上精彩的一笔。

沿街人视图

北京市民政工业总公司养老产业项目

1. 项目概况

（1）建设地点：北京市。

（2）建设规模：用地面积 23774.565m²；建筑面积 92485.4m²。

（3）建设情况：建设中。

（4）床位数：1300 张。

2. 项目简介

北京市民政工业总公司养老产业项目以中国传统的庭院为中心，用建筑包围庭院，形成以外部空间为中心的建筑组合，空间层次丰富，仿佛是快节奏城市背景下的一处净土，为老年人提供全面、专业的医疗养老服务。

本项目规划采用 L 形布局，建筑沿场地展开，底层空间采用连续的建筑空间加以联系，形成院落式的空间格局，实现活跃多元的现代养老空间。项目床位数约 1300 张，设有二级医院，基础功能完备，可以满足养老院及周边区域日常医疗服务的需求。目前本项目是北京市规模最大的在建养老院。

沿街立面效果图

湖北省医养康复中心（示范）项目

1. 项目概况

（1）建设地点：武汉市洪山区卓刀泉南路西侧。

（2）建设规模：总用地面积 60972m²，其中东侧地块位于卓刀泉南路西侧，净用地面积 36797m²，西侧净用地面积 8337m²。拟建地上建筑面积 85882m²，地下建筑面积 35118m²。

（3）建设情况：未建成。

（4）床位数：1200 张。

（5）业主单位：湖北省荣军医院。

2. 项目简介

该项目综合考虑三级康复医院与养老院的面积标准、床位标准、规范标准、建设标准，是在 12.1 万平方米的建筑面积内设计 1200 张病床的医养融合项目，既满足医疗功能的需求，又满足养老的舒适性需求。同时通过对全国 7 个省、36 个项目细致的调查与研究，探索出标准护理比例、50 人 1 组的标准活动空间设置，以及其他适老性设施的设计参数。

该项目地块土方量较大，设计通过抬高建筑正负零标高高度，减少土方量约 10 万立方米，同时营造了更好的城市景观、观湖景观及建筑的私密性环境景观。

沿湖效果图

参考文献

[1] 许晓华.雷神山医院建设亲历者讲述"中国速度"：动工到收治病人仅用 13 天 [EB/OL]. （2020-05-04）[2020-05-05].http://health.people.com.cn/n1/2020/0504/c14739-31696929.html.

[2] 新华网.习近平向 2019 中国国际数字经济博览会致贺信 [EB/OL].（2019-10-11）[2020-05-05].http://www.xinhuanet.com/politics/leaders/2019-10/11/c_1125091565.htm.

[3] 新华网.国家主席习近平发表二〇一九年新年贺词 [EB/OL].（2018-12-31）[2020-05-05].http://www.xinhuanet.com/politics/2018-12/31/c_1123931806.htm.

[4] 仇争艳，周里，刘丰祺.武汉雷神山医院规划策略纪实 [J].华中建筑，2020，38（4）：28-31.

[5] 张姗姗，朱丽玮.医院建筑装配式建设趋势与空间模块化实现 [J].城市建筑，2017（13）：30-32.

[6] 中华人民共和国住房和城乡建设部.钢管约束混凝土结构技术标准：JGJ/T 471—2019[S].北京：中国建筑工业出版社，2019.

[7] 江苏省住房和城乡建设厅.施工现场装配式轻钢结构活动板房技术规程：DGJ32/J54—2016[S].南京：江苏凤凰科学技术出版社，2016.

[8] 黄锡璆.小汤山医院二部工程概述 [J].工程建设与设计，2003（6）：3-6.

[9] 秦晓梅，胡颖慧，危忠，等.方舱医院给水排水及消防系统设计——以江夏大花山户外运动中心乒羽馆改造工程为例 [J].给水排水，2020，56（4）：32-35.

[10] 中华人民共和国国家卫生健康委员会办公厅，中华人民共和国住房和城乡建设部办公厅.关于印发新型冠状病毒肺炎应急救治设施设计导则（试行）的通知：国卫办规划函〔2020〕111 号 [A/OL].（2020-02-08）[2020-05-05].http://www.gov.cn/zhengce/zhengceku/2020-02/11/content_5477301.htm.

[11] 王涛，吴平，秦晓梅，等.方舱医院污水收集处理系统现状及对策分析 [J].给水排水，2020，56（5）：22-26.

[12] 中华人民共和国生态环境部办公厅.关于做好新型冠状病毒感染的肺炎疫情医疗污水和城镇污水监管工作的通知：环办水体函〔2020〕52 号 [A/OL].（2020-02-01）[2020-05-05].http://www.mee.gov.cn/xxgk2018/xxgk/xxgk06/202002/t20200201_761163.html.

[13] 武汉市生态环境局，武汉市卫生健康委员会，武汉市水务局，等.关于做好全市方舱医院医疗污水处理有关工作的紧急通知 [A].（2020-02-04）[2020-05-05].

[14] 应急管理部消防救援局.发热病患集中收治临时医院防火技术要求 [A/OL].（2020-01-26）[2020-05-05].http://119.guizhou.gov.cn/xwdt/tzgg/202002/t20200205_47234210.html.

[15] 湖北省住房和城乡建设厅.关于印发《方舱医院设计和改建的有关技术要求》的通知：鄂建函〔2020〕22 号 [A/OL].（2020-02-06）[2020-05-05].http://www.hubei.gov.cn/hbfb/bmdt/202002/t20200206_2020167.shtml.

[16] 中华人民共和国住房和城乡建设部，中华人民共和国国家质量监督检验检疫总局.传染病医院建筑设计规范：GB 50849—2014[S].北京：中国计划出版社，2015.

[17] 中华人民共和国住房和城乡建设部.医疗建筑电气设计规范：JGJ 312—2013[S].北京：中国建筑工业出版社，2014.

[18] 刘荔，张毅，付林志，等.热分层环境人际间飞沫传染风险与对策研究 [J].暖通空调，2020，50（6）：19-25.

[19] 湖北省住建厅"方舱医院设计和改建的有关技术要求"2020.02.

[20] 江苏苏净科技有限公司，天津市龙川净化工程有限公司，中天道成（苏州）洁净技术有限公司，等.医院负压隔离病房环境控制要求：GB/T 35428—2017[S].北京：中国标准出版社，2017.

[21] 钱华，郑晓红，张学军.呼吸道传染病空气传播的感染概率的预测模型 [J].东南大学学报，2012.42（3）：468-472.

[22] 郝吉明，马广大，王书肖.大气污染控制工程 [M].北京：高等教育出版社，2010：87-107.

[23] GRIEVESM. Digital twin : manufacturing excellence through virtual factory replication[M].[s.l.] : [s.n.]，2014.

[24] 张玉彬，赵奕华，李迁，等.基于 BIM 竣工模型的医院智慧运维系统集成研究 [J].工程管理学报，2019，33（2）：141-146.

[25] 王凯.基于互联网 + BIM 的智慧医院的展望与思考 [J].土木建筑工程信息技术，2017，9（3）：94-97.

[26] 吴杰.格子 Boltzmann 方法及其应用研究 [D].南京：南京航空航天大学，2005.

[27] TRAPANI G, HOLMAN D M, BRIONNAUD R.Non-linear fluid-structure interaction using a partitioned lattice boltzmann-FEA approach[C]//46th AIAA Fluid Dynamics Conference.2016.

[28] 张慎，尹鹏飞.基于 Rhino+Grasshopper 的异形曲面结构参数化建模研究 [J].土木建筑工程信息技术，2015，7（5）：102-106.

[29] JIANG Y, ZHAO B, LIX F, et al. Investigating a safe ventilation rate for the prevention of indoor SARS transmission : an attempt based on simulation approach[J].Building Simulation，2009，2（4）：281-289.

跋

经过几个月的精心准备，《数字化应急医院设计及建造技术》终于付梓成册，与读者见面了。在此特别感谢中国工程院王辰院士、中南设计集团张柏青董事长在本书编撰过程中给予的关心和指导。

突如其来的疫情，让武汉这座城市的人们经历了很多。中南院和武汉这座英雄的城市一起，以战士的姿态顽强地冲锋在战"疫"前线。作为公司董事长，我为我们的设计师感到骄傲和自豪。接到雷神山医院、方舱医院设计任务时，正是武汉市疫情形势最为严峻的时候。让人非常感动的是，我们的设计师主动请缨、前赴后继，没有因邻居、朋友已有人感染而退缩和犹疑，冒着被感染的风险，第一时间赶到项目现场，通宵达旦开展抢建设计。

回想数十年前，中南院的老前辈们支援国家三线建设、修建二汽厂房、驰援唐山大地震后城市重建、扎根沿海城市投身改革开放建设，设计师的画笔始终围绕着国家建设的需要挥毫。疫情发生，我们新一辈的员工接过前辈的旗帜，传承可贵的奋斗、奉献精神，在抗疫中彰显大义。正是怀着强烈的责任感和使命感，怀着对生命的无限敬畏，他们义无反顾、勇往直前，战斗在项目建设的最前线，出色地完成了一个又一个抗疫抢建项目！

历史上历次重大公共卫生事件，都是对城市建设管理的"大考"。本次新型冠状病毒来袭，给城市防灾救灾体系建设带来考验。作为城市建设管理的重要参与者，建筑设计者责无旁贷。我们既要立足城市功能定位和总体规划理念，又要积极研究城市功能配套设施的完善；我们既要引导人们树立绿色健康的生活理念，又要着力为人们提供舒适、通风、环保的居住空间。我们谨以此书抛砖引玉，希望能够引发业内人士的广泛研究、交流、讨论，形成新理念下的技术体系。后疫情时代，我们呼吁人居环境安全健康成为更受关注的主题，我们欢迎国内外业内同仁与我们开展讨论和交流，期待共同形成更多的成果并进行推广应用，为改善人居环境做出共同的努力。

经过本书编写组各工作团队的共同努力，本书尽可能翔实地记录雷神山医院、方舱医院设计建设过程中各专业总结的设计要点和思考，尽可能全面地分享抗疫应急工程设计建设技术资源，我们希望它能给有需要的地区政府管理部门、科研机构和同行设计单位提供有益的借鉴。

本书编写组各团队，特别是技术组、宣传组和翻译组，他们在很短的时间内完成本书策划和编撰工作，付出了辛苦的努力，限于篇幅，不逐一列出他们的名字，在这里一并向他们表示感谢！同样限于篇幅，对于编委会中参加多项工作的人员，其姓名在编委会名单中各部分不重复出现。

衷心感谢各位读者给予的关注，希望本书的出版能给大家以启发，如果本书所述的观点和技术有误，敬请指正。

同在地球村，让我们携手同心，共同期待全球战"疫"胜利的那一天！

李　霆

全国工程勘察设计大师

中南建筑设计院股份有限公司董事长、总工程师

图书在版编目（CIP）数据

数字化应急医院设计及建造技术 = Design and
Construction Technology of Digital Emergency
Hospital / 中南建筑设计院股份有限公司编著 . —北京：
中国建筑工业出版社，2022.5
ISBN 978-7-112-27962-3

Ⅰ.①数…　Ⅱ.①中…　Ⅲ.①数字技术—应用—传染
病医院—建筑设计　Ⅳ.① TU246.1-39

中国版本图书馆 CIP 数据核字（2022）第 174362 号

责任编辑：刘　静　陆新之
书籍设计：康　羽
责任校对：赵　菲

数字化应急医院设计及建造技术
Design and Construction Technology of Digital Emergency Hospital
中南建筑设计院股份有限公司　编著
*
中国建筑工业出版社出版、发行（北京海淀三里河路 9 号）
各地新华书店、建筑书店经销
北京雅盈中佳图文设计公司制版
北京雅昌艺术印刷有限公司印刷
*
开本：850 毫米 ×1168 毫米　1/16　印张：15¹/₂　字数：357 千字
2022 年 11 月第一版　2022 年 11 月第一次印刷
定价：**199.00** 元
ISBN 978-7-112-27962-3
　　　（38983）